Conquest of Invisible Enemies

Conquest of Invisible Enemies

Enemies

A Human History of Antiviral Drugs

JIE JACK LI

OXFORD
UNIVERSITY PRESS

OXFORD
UNIVERSITY PRESS

Oxford University Press is a department of the University of Oxford. It furthers
the University's objective of excellence in research, scholarship, and education
by publishing worldwide. Oxford is a registered trade mark of Oxford University
Press in the UK and certain other countries.

Published in the United States of America by Oxford University Press
198 Madison Avenue, New York, NY 10016, United States of America.

Library of Congress Cataloging-in-Publication Data
Names: Li, Jie Jack, author.
Title: Conquest of invisible enemies : a human history of antiviral drugs / Jie Jack Li.
Description: New York, NY : Oxford University Press, 2022. |
Includes bibliographical references and index.
Identifiers: LCCN 2022029911 (print) | LCCN 2022029912 (ebook) |
ISBN 9780197609859 (hardback) | ISBN 9780197609873 | ISBN 9780197609866 |
ISBN 9780197609880 | ISBN 9780197609873 (epub)
Subjects: LCSH: Antiviral agents.
Classification: LCC RM411 .L55 2022 (print) | LCC RM411 (ebook) |
DDC 615.7/924—dc23/eng/20220802
LC record available at https://lccn.loc.gov/2022029911
LC ebook record available at https://lccn.loc.gov/2022029912

DOI: 10.1093/oso/9780197609859.001.0001

1 3 5 7 9 8 6 4 2

Printed by Sheridan Books, Inc., United States of America

To Vivien and Alexandra

Contents

Preface   xi

1. Viruses Shaping History and the Discovery of Viruses   1
   1.1 Viruses Shaping History   1
       1.1.1 Tulipomania: Bubbles of bulbs   2
       1.1.2 The fall of old Mexico and Peru   5
       1.1.3 Succession of the throne   10
       1.1.4 The Louisiana Purchase   15
   1.2 Discovery of Viruses   19
       1.2.1 The germ theory   19
       1.2.2 Discovery of viruses   22
       1.2.3 Cancer-causing viruses   25
       1.2.4 Discovery of retroviruses   32

2. HIV/AIDS Drugs: Transforming a Death Sentence
   into a Chronic Disease   37
   2.1 Discovery of the Virus   38
       2.1.1 A death sentence   38
       2.1.2 Discovery of HIV and the great scientific controversy   40
       2.1.3 The Nobel Prize   50
   2.2 The First Ray of Hope   56
       2.2.1 Nucleoside antiviral drugs   56
       2.2.2 Non-nucleoside reverse transcriptase inhibitors   65
   2.3 HIV Protease Inhibitors   69
       2.3.1 HIV protease   69
       2.3.2 First-generation HIV protease inhibitors   71
       2.3.3 Second-generation HIV protease inhibitors   77
   2.4 HIV Integrase Inhibitors   80
   2.5 HIV Entry Inhibitors   83
       2.5.1 HIV fusion inhibitors   84
       2.5.2 CCR5 inhibitors   86
   2.6 The Road to Eradication   89

3. Hepatitis Viruses   91
  3.1 Hepatitis A Virus   92
    3.1.1 Discovery of the hepatitis A virus   93
    3.1.2 Hepatitis A vaccines   95
  3.2 Hepatitis B Virus   98
    3.2.1 Discovery of the hepatitis B virus   99
    3.2.2 Hepatitis B vaccines   106
    3.2.3 Hepatitis B drugs   112
  3.3 Hepatitis C Virus   125
    3.3.1 Discovery of hepatitis C virus   126
    3.3.2 Hepatitis C virus NS3/4A serine protease inhibitors 130
    3.3.3 Hepatitis C virus NS5A protein inhibitors 139
    3.3.4 Hepatitis C virus NS5B polymerase inhibitors 143
  3.4 A Triumph of Modern Medicine   151

4. Influenza: A Perennial Killer   154
  4.1 Influenza, A Perennial Killer   155
  4.2 A Virus that Changed the World:
    The 1918 Spanish Flu Pandemic   157
  4.3 Discovery of the Influenza Virus   161
  4.4 Influenza Vaccines   165
    4.4.1 History of vaccination   165
    4.4.2 History of influenza vaccines   168
    4.4.3 Influenza vaccines today   170
  4.5 Influenza Drugs   172
    4.5.1 M2 inhibitors   172
    4.5.2 Influenza virus neuraminidase inhibitors   175
    4.5.3 Cap-dependent endonuclease inhibitor   184

5. Coronaviruses   187
  5.1 Coronavirus   188
    5.1.1 What is a coronavirus?   188
    5.1.2 SARS and MERS   189
    5.1.3 SARS-CoV-2   190
  5.2 COVID-19 Vaccines   192
    5.2.1 Pfizer/BioNTech's mRNA vaccine   193
    5.2.2 Johnson & Johnson's viral vector vaccine   196

5.3 COVID-19 Drugs   202
    5.3.1 Remdesivir (Veklury)   202
    5.3.2 Dexamethasone (Decadron)   211

6. Closing Remarks   217

Appendix: Chemical Structures of the Drugs   219
Notes   235
Index   249

Preface

COVID-19 is wreaking havoc around the world. Many of us have wondered:

- Where is the virus coming from?
- What is a coronavirus?
- Where did the treatment drugs come from?
- How do these drugs work?

As a scientist, it is unfathomable to me that so much anti-science sentiment exists in America. Even party affiliation seems able to determine if one believes in science. More than ever, science matters. I decided to write a book about viruses and antiviral drugs because, after all, history can teach us invaluable lessons.

In Chapter 1, I highlight the role that viruses have played in human history and the discovery of viruses. In Chapters 2 to Chapter 5, I cover HIV, hepatitis viruses, influenza virus, and coronavirus. Each chapter starts with a discussion about the disease that the virus causes, followed by the discovery of the virus, vaccines, and antiviral drugs. I hope that healthcare workers, as well as laypeople, can learn from the history of viruses, diseases, vaccines, and antiviral drugs. Unlike most scientific books, this book attempts to focus on the human aspect of discovery. After all, drug discovery and development are carried out by humans for humans. Consequently, scientists display just as much humanity as people in any other occupations.

I welcome your critique on this important topic. Please send your comments directly to lijiejackli@hotmail.com.

Jie Jack Li
San Mateo, CA
June 1, 2021

1

Viruses Shaping History and the Discovery of Viruses

"Virus is a piece of nucleic acid surrounded by bad news."
—*Peter B. Medawar, 1960 Nobel Laureate*
for Physiology or Medicines

"Germs" in our daily vernacular imply either bacteria or viruses, both of which cause terrible infectious diseases. But they are profoundly different in many aspects. In size alone, a bacterium could be fifty to two hundred times larger than a virus. If a bacterium were the size of a man, a virus would be about the size of an arm. A cell is one million times larger than a virus! Antony van Leeuwenhoek, a Dutch draper merchant, observed bacteria using microscopes in the 1670s. The simple one-lens microscopes that he ground himself only had a three-hundred-fold magnification. But viruses are so small that they could not be "seen" until after the electron microscope was invented in 1931. The first electron microscope had less than a one-thousand-fold magnification, but today's electron microscopes have millions-fold of magnification.

1.1 Viruses Shaping History

Even though viruses were invisible to man until the 1930s, their impact has been felt throughout human history. Here, I share some

Conquest of Invisible Enemies. Jie Jack Li, Oxford University Press. © Oxford University Press 2022.
DOI: 10.1093/oso/9780197609859.003.0001

historical vignettes associated with viruses and how they drastically altered the trajectories of history. One of them, known as tulipomania, caused the earliest speculative financial bubble.

1.1.1 Tulipomania: Bubbles of bulbs

The seventeenth century was the Golden Age of the Dutch Republic. From 1634 to 1637, both well-to-do rich regents and merchants and also poor artisans seemed to be seized by a craze, an obsession, or more aptly, a mania: tulipomania!

Wild tulips flourished among the foothills and valleys of Tian Shan (天山), the Celestial Mountains that run along China's western border neighboring both Russia and Afghanistan.[1] By the eleventh century, Turkish nomads had brought the wildflower to Persia, most likely along the Silk Road. Admired for the vivid sheen of the flower's petals, wild tulips were gradually planted in gardens and domesticized in the Ottoman Empire. In Islamic literature, tulips were worshiped as a symbol of holiness and love during the fifteenth and sixteenth centuries. Slowly but surely, the oriental flower migrated westward. Its wide distribution was made easy by its portability as either seeds or bulbs, and the latter was the preferred form because seeds took years to bear flowers. In 1576, Dutch botanist Carolus Clusius, the father of the tulip, first reported viral flower color-breaking. He described a variegation of the flower color in striped and flamed patterns in combination with a weakening of the tulip plants, eventually leading to the loss of varieties. This was the oldest documented plant disease.[2]

By the early seventeenth century, the subtle elegance, novelty, and rarity of tulips made the flowers a symbol of wealth and good stature in Europe. In Paris, already the center of European fashion, a cut tulip flower became a fashionable adornment to the ladies' cleavage in high society. While fashions came and went, it was the Dutch people of all classes who became seriously infatuated with tulips. Connoisseurs in the Netherlands paid exorbitant prices for rare tulips. In the summer of 1633, a house in the port town of Hoorn was sold to pay for three rare tulips. While the scarcity of some tulips always fetched high prices, things truly went awry at the end of 1636 and the beginning of 1637. The popularity of broken tulips made trading tulips extremely lucrative. For several months, trading tulips—any tulips—could always fetch an exorbitant profit one way or the other. Even normally worthless, plain tulip bulbs were sold for hundreds of guilders. Almost everyone in the Netherlands got into the business of trading bulbs. Sometimes a bulb changed hands more than ten times in a single day! At the height of the tulip craze, three thousand guilders (the annual income of a well-off merchant) were needed to secure a single bulb of the rarest variety, *Semper Augustus*, revered as the empress of tulips. Sadly, like all economic bubbles, the tulip bulb craze crashed precipitously when the market for tulips simply ceased to exist and there was suddenly no demand for the bulbs any longer. After the burst of the bubble, the bulbs were worth only one percent of peak prices and numerous tulip speculators in Holland went bankrupt. Economists consider that incident the first financial bubble in history. Human greed has no bounds. The dot-com bubble in the late 1990s and the housing bubble in 2006 in the United States are just more spectacular examples in recent memories.

In 1850, the famous French author Alexandre Dumas published the novel *Black Tulip*. The story took place in 1672 in the city of Haarlem, where a price of one hundred thousand guilders was offered to the gardener who could produce a black tulip. Fiction aside, the most desirable and sought-after tulips were actually *broken tulips*. The petals of broken tulips had beautiful, variegated color patterns that broke the solid color of a normal tulip, hence the name *tulip breaking*. It has been described as stripes, streaks, feathering, or flames of different colors on petals (see Figure 1.1). Broken tulips were also known as Rembrandt tulips, although the Master rarely painted them. With spectacular patterns of color striations,

Fig. 1.1 Tulip © Netherland Post

broken tulips were highly appreciated and served as the catalyst for the tulipomania. Such virus-infected tulips were not healthy, and so they did not breed true. The color-breaking pattern suddenly died out after having been a craze in Holland for four years.

In the 1930s, Dorothy Cayley and Charles Mackay at the John Innes Horticulture Institution in London discovered that variegation of tulips was caused by the *tulip breaking virus*. By permitting aphids to feed on broken bulbs (virus-infected, which thus would give broken tulips) and then on breeders (not infected, which thus would not give broken tulips), they were able to show that the breeder bulbs visited by the aphids broke twice as often as control samples. The British duo simultaneously proved that the disease was caused by a virus and demonstrated the mechanism whereby it was transmitted from one tulip to another.[3]

The tulip-breaking virus, *tulip mosaic virus*, is a ribonucleic acid (RNA) plant potyvirus. Indeed, the great majority of plant viruses are RNA viruses; only about seventeen percent of viruses infecting plants are deoxyribonucleic acid (DNA) viruses. On a molecular level, flower color is largely determined by the different accumulation of plant pigments made up by a class of colorful molecules—anthocyanins. They are flavonoids that show different colors under the sun, depending on the molecular substituents. For instance, the molecule 6′-deoxychalcone is yellow, and glucoside-attached anthocyanins are red, dull grey, magenta, purple, etc. Virus-induced tulip breaking involved modifications of a class of enzymes that participate in the anthocyanin biosynthesis pathway. Those enzymes include chalcone reductase, anthocyanidin synthase, chalcone synthase, dihydroflavonol 4-reductase, 3′-glucosyltransferase, and rhamnosyltransferase, among others.[4]

Today, the tulip trade is a $3 billion business and is especially important to the Dutch economy. The virus that caused the fabled broken tulips no longer exists. Destruction of the tulip-breaking virus was the florists' equivalent of the elimination of smallpox.[5] Novel varieties of new genera of tulips are now created via hybridization.

The tulip breaking virus has been touted as an artistic agent and it is also known as the Rembrandt virus because of the beautiful, spectacular variegation that it brought on tulip petals. Unfortunately, not all viruses are as benevolent as the tulip-breaking virus. Many viruses are deadly. One of the most devastating of all is the smallpox virus.

1.1.2 The fall of old Mexico and Peru

In 1519, driven by a greed for gold, thirty-four-year-old Hernán Cortés and his Spanish conquistadores embarked on an expedition to Mexico. In the summer of 1521, he led approximately one thousand Castilians and their Indian allies and he subjugated the Aztec empire with hundreds of thousands of warriors after laying a siege of Tenochtitlan, Mexico's then capital. Cortés's conquest was aided by technological, psychological, and political advantages. But perhaps his greatest advantage was infectious diseases in general, and

smallpox in particular, which played a decisive role during the conflict between the Old World and the New.

In terms of technology, the Spaniards sailed on great ships and
brigantines. They were equipped with armaments of the Iron
Age: steel swords, steel lances, crossbows, harquebuses (match-
lock firearms), and canons. Their thunderous muskets and artil-
lery might be considered primitive by today's standards, but they
were psychologically terrifying at the time to the indigenous people
in South America. Even horses, used for the first time in battles in
Americas, were frightening to the natives as they never had seen
the animal before. In stark contrast, the Aztec's weaponries were
still of the Bronze Age: copper and bronze clubs, bows and arrows,
and stone slings. These weapons were suitable to injure, not to kill,
fitting the purposes of capturing prisoners of war for the human
sacrifice rituals that were popular in the Aztec culture.

Cortés's invasion of Mexico, like all similar missions, was invariably accompanied by Catholic friars or priests. Inspired by reli-
gious fervor, some of the conquistadores felt that they were on the
righteous side of the clash even though plundering was the overwhelming goal for all the conquistadores. They considered the
natives savage, and they found the ritual of human sacrifice and
cannibalism barbaric, notwithstanding their own extreme sav-
agery of brutal massacres of the Indians at Cholula, Teapeca, and
Tenochtitlan. In the early part of the sixteenth century, South
America was divided, by papal decree, between Spain and Portugal
even though millions of indigenous people had already lived there
for millions of years. The cruel conquistadores carried out the most
despicable atrocities in plundering and murdering the indigenous
people, a far cry from true Christian values. Indeed, throughout
human history, many crimes have been committed in the name
of religions. Inversely, the superstitious Mexicans mistook the
bearded, white-skinned Spaniards as the incarnation of a lost
lord from their mythical legends and they offered little resistance
physically or psychologically. Cortés proved to be a consummate

politician and diplomat. He cultivated alliance with the Totonacs, the Tlaxcalans, the Chalca, and even the Cholulans and the Huexotzinca, which afforded him thousands of indigenous men as soldiers and auxiliary service members against their common adversary, the Aztecs.[6]

But there is no denying that infectious diseases, especially smallpox, played a profound role in their astounding conquest. Smallpox killed far more Aztec citizens in deathbed than conquistadores' guns and swords on the battlefield.

Smallpox is an infectious disease caused by the *variola virus*, which is a DNA virus. It is a member of the genus *orthopoxvirus* in the family of *Poxviridae*. It might have emerged three thousand to four thousand years ago, and it most likely evolved from cattle (cowpox) or other livestock with related pox viruses. At the time, with the exception of tuberculosis and influenza, a single infection by most viruses induced prolonged, often life-long, immunity, which worked in favor of the Spaniards from the Old World. In contrast, these same infections were likely to kill a high proportion of human population in the New World who fell sick and lacked any previous exposure to the contagions.

There were recorded smallpox epidemics in China and Europe as early as in the first century. The invasion of the Huns and Genghis Khan's cavalry probably helped spread smallpox across Eurasia. Around the twelfth century, the Crusades contributed vastly to the dissemination of smallpox in Europe.[7] The first reported case of smallpox in the New World occurred in 1507 on an island in the West Indies; presumably it was introduced by sailors from Spain. It became an epidemic in Hispaniola (the island upon which Haiti and Dominique Republic reside today) in 1518, killing one-third to one-half of the native population, who were already suffering from the forced hard labor of mining and agriculture. But few Spaniards perished. Whenever a new disease is introduced into a naïve population, the results can be devastating. But lifelong immunity followed recovery from infection of the variola virus. At that time, smallpox was endemic in Asia, Africa, and Europe, where the Spaniards were exposed to the pathogen at young ages and developed immunity. To make up for the severe depopulation and resulting labor shortage created by the deaths of the indigenous people, the Spaniards began shipping African slaves, and these slaves brought more smallpox infections. Germs—not just people—were also active participants of the collision between the New World and the Old. The Spaniards were farmers with domesticated animals as livestock that served as reservoir of viruses. They developed a certain immunity for the contagions. The Aztecs were hunters-gatherers, a lifestyle that kept them away from germ-ridden animals. They had no inherited or acquired immunity against the European infections.

At the end of 1519, the smallpox outbreak reached Cuba when Cortés sailed from Santiago to Mexico after two prior failed expeditions led by Hernández de Córdoba and Juan de Grijalva, respectively.[8] Escaping smallpox surely added more incentives to Cortés's abrupt departure from Cuba. A smallpox outbreak in Mexico began with an African slave in Cortés's camp in April 1520, and it slowly spread inland. It moved from place to place at the deliberate pace associated with the disease and its pre-infective period of more than two weeks. The incubation period of a viriola virus infection is two weeks and it takes a few days for the virus to reach the point of infectiveness. In five months, from May to September, the epidemic moved across some 150 miles and was finally exhausted in January 1521.[9] The indigenous Mexicans had never encountered smallpox before and so the naïve population had no immunity. Town after town became depopulated. In many streets there was no way of collecting corpses. Particularly, the Aztec high society died at an even higher rate than their subjects, probably because they were in more close contacts with the Castilians. Because most Spaniards were impervious to the disease, the Aztec indigenous people naturally thought that the visitation of smallpox was a punishment by angry Gods for their blasphemy. The fact that most Europeans were immune must be a sign that they were the Gods' favorites or, perhaps, they were an incarnation of Gods. Some exaggerated that the indigenous people died so easily that the bare look and smell of a Spaniard caused them to give up the ghosts.[10]

There are two types of variola viruses. One is *variola major*, which kills one in three people infected. The other type is *variola minor*, which is a naturally occurring variant that is lethal to only one in one hundred people infected. The smallpox major virus is responsible for wiping out a large swath of humanity throughout history. There is no doubt that the virus that ravaged Mexico in 1519 was the more lethal *variola major*. In 1520, Mexico had a population of twenty million, but it had plummeted to 1.6 million by 1618 under the incessant onslaught of infectious diseases, among other factors.

A short decade later, Pizarro's easy conquest of Peru in 1532 was almost a repeat of Cortés's exploits in Mexico. Francisco Pizarro was actually Hernán Cortés's second cousin, although he was seven years older than Cortés. The illiterate Pizarro made the third expedition to Peru in 1532 with a small band of 167 Spanish conquistadors. The Inca Empire was much weakened at the time already because the smallpox virus had spread overland across South America after its arrival with Spanish settlers in Panama and Columbia. In 1526, the Inca emperor Huayna Capac died of "pestilence" during a mysterious plague, which was almost certainly smallpox. Even though his son, Atahualpa, emerged as a victor during a bloody campaign against his brother, Huascar, the Inca Empire was severely weakened by the violent civil war and the European smallpox.

At Cajamarca, the two worlds met, and two civilizations collided. As one of the most extraordinary and astonishing events in history unfolded, Pizarro captured the Inca Sun-God Atahualpa within a few minutes after the two leaders first set eyes on each other. That a band of 168 conquistadores was victorious over an emperor with eighty thousand warriors in the vicinity was truly astounding! In the ensuing years, Pizarro, his four half-brothers, and their fellow conquistadores brutally massacred thousands of the Incas in an attempt to seize gold and silver and to subjugate the natives. It was perhaps a poetic justice that Pizarro was murdered by another faction of the conquistadores who did not receive their "fair share" of the spoils.[11]

It is not exaggerating to say that the smallpox virus worked in tandem with the conquistadors in decimating the indigenous populations in Americas brought by European colonists.[12] With *guns, germs,* and *steel* as their accomplices, the two avaricious and brutal Castilian cousins—Cortés and Pizarro—with small bands of conquistadors overpowered two of the largest empires of the New World.

The depopulation wiped out ninety-five of the original inhabitants in the Americas. More than three million Amerindians perished by the inadvertent spread of measles, smallpox, and influenza. More deplorably, smallpox was used as a biological warfare during the French and Indian war (1754–1763). To capture Fort Pitt (today's Pittsburg), the British commander Jeffrey Amherst suggested to Colonel Bouquest that they "send the smallpox among those disaffected tribes of Indians . . . . We must on this occasion use every strategy in our power to reduce them."[13] Captain Ecuyer of the Royal Americans gave to the tribes two blankets and a handkerchief smeared with ground smallpox scabs from the smallpox hospital. The fort was easily captured after smallpox raged through the tribes of Ohio.

1.1.3 Succession of the throne

Smallpox was also a key factor in deciding who took over the throne of the Qing Dynasty (1644–1911) in China.

The Qing Dynasty was founded in 1644 when the Manchus conquered Ming China. An ethnic minority, the Manchus were a half-nomadic and half-agricultural people who lived around

China's vast northeastern borders, Manchuria. The climatic conditions and lifestyle enabled most of the Manchu population to keep their distance from smallpox contagion; as a result, they had no immunity against the virus. After the Manchus entered China, living in the crowded Forbidden City, the conquerors were in close contact with the local Han Chinese. In China proper, smallpox was endemic and people had been exposed to the disease thus giving immunity to many already. For the Chinese majority Han ethnicity, long-term exposure to smallpox led to early onset of infection. Small children tended to catch the disease and had a better chance to survive unscathed. At the time, once contracted, the Manchus, who were naïve to the contagion, almost invariably died as they had no immunity against smallpox virus infection. As a result, they dreaded smallpox more than any other disease. Their healthy level of respect for smallpox proved to be pivotal to their conquest and their rule of China. As an aside, it is always good that a historian knows some science. Writing Chinese history, historians always express their surprise that the Han Chinese tended to survive smallpox better as a population, but eighty to ninety percent of the infected Manchus died of smallpox "mysteriously." If the historian–authors knew a little bit about immunology, they would have known that there was nothing "mysterious" about this phenomenon.

The situation of the Qing Dynasty was not so dissimilar to Native Americans before Columbus' "discovery" of the new continent in 1492. The North America population had never had smallpox before the Europeans showed up. The "Indians" had no immunity against the new viral pathogen, which was one of the reasons why the white colonists chose to import slaves from Africa. They purposely targeted Africans from the regions where smallpox was endemic.

Back to Chinese history, the smallpox virus became a major factor shaping politics, the military, the economy, and the diplomacy of the Qing Dynasty. They learned that anyone who survived smallpox would never be affected again. Thus, contraction of smallpox was a watershed moment in life of the Manchus.

Even before conquering China, the Manchus already established a smallpox tracking agency as early as in 1622, which was definitely the first in Chinese history. The initiative started with its army, the "Eight Banners," but expanded to all subjects as their territory slowly encroached upon the warmer China proper. Once a person was suspected to have contracted smallpox, that person was immediately sent to a separate location for quarantine.

For the royal family, once there was a smallpox outbreak, the entire royal family fled to the countryside where they built many quarantine shelters and stayed there until the epidemic subsided. At Emperor Abahai's funeral in 1643, royal members who had not yet had smallpox and those had been affected congregated separately. It was the fear of smallpox that kept these two groups apart during such an important occasion. This might be one of the earliest examples of "social distancing." Active management of the threat of smallpox enabled a feat achieved by the Manchus opposite to Cortés and his Spanish conquistadors. For the Qing dynasty, it was the conquerors who lacked immunity to smallpox, but triumphed over the conquered populace who had certain immunity already.

Emperor Shunzhi (順治, 1638–1661) ascended the throne in 1644 when he was six years old, becoming the first Manchu emperor to live in China where smallpox was more common. Young, naïve to the smallpox virus, and living in the largest city Beijing, Shunzhi was especially vulnerable to the contagion. During his seventeen-year reign, there were at least nine recorded smallpox outbreaks. Shunzhi always confined himself to the many quarantine shelters that were built for him. Once he even escaped to the Northeastern border to outrun the infectious disease.

Ironically, despite his great fear of and immense efforts to avoid smallpox, Shunzhi died of that very disease. Before his death, he decreed that those who had not had smallpox could not succeed to the throne. Kangxi (康熙) became the next Emperor when he was only seven years old. One of the major reasons why he was chosen was because he had survived smallpox already and was thus less likely to die if he were infected again. Kangxi promoted the practice *variolation* of smallpox throughout his reign. Variolation was a sophisticated technique developed in China in the 1500s. Although the "learned scholars" at the time sneered at the "absurd" idea, the imperial edict prevailed and variolation was implemented in China.[14] Emperor Kangxi ruled China for sixty years (1662–1722) and it was apparently a period of great prosperity when Chinese economy flourished.

China was not the only country inflicted by smallpox. In 1562, the twenty-nine-year-old Queen Elizabeth I of England was inflicted with smallpox. Her face was scarred but not badly disfigured, which was probably why the Virgin Queen covered her pockmarks with heavy white makeup. A number of European monarchs and heirs also succumbed to the deadly smallpox virus around that period of time, including:

- Queen Mary II of England (1694),
- Duke of Gloucester (the son of Queen Anne, 1700),
- Austrian Emperor Joseph I (1711),
- King Louis XV of France (1794), and
- Tsar Peter II of Russia (1739).

The death of Tsar Peter II, the only grandson of Peter the Great, was immensely consequential. Although smallpox was relatively rare in Russia compared to Western Europe, he died of smallpox in 1730 at a young age fifteen, on his wedding day, after only three years on the throne. His untimely death profoundly altered the trajectory of Russian history. Lacking a male heir, the house of Romanovs enthroned Anna as empress; she ruled for ten years (1730–1740). In 1741 Elizabeth, one of Peter the Great's two daughters, was coronated. Having lost her fiancée, a prince of Holstein–Gottorp in Germany, to smallpox right before their wedding day,

Tsarina Elizabeth I never married and had no heir. She decreed her nephew, Peter, would take over in 1762 when she passed away.

Tsar Peter III suffered a severe case of smallpox at the age of twenty-five that left his face scared by "hideous" pocks and possibly destroyed his confidence. Growing up in Holstein and fiercely pro-Prussian, his reign was not popular and was short-lived. Within only six months, he was overthrown by patriots who were led by his wife, Ekaterina (Catherine), in 1762. Tsarina Catherine II, later known as Catherine the Great, was a German princess. After ascending the throne, she became a staunch promoter of smallpox vaccination, inoculating herself and her son, the Grand Duke Paul, in 1798 by English doctor Thomas Dimsdale[15] (see Figure 1.2). It was impressive that she, risking her life, was the first to undergo inoculation in the Russian empire, setting a personal example to autocrats at the court and to all citizens, military men, and the clergy. The empress signed an edict on obligatory vaccination. Smallpox houses were opened all around Russia. On the eve of the Russo–Turkish war (1768–1774), Ottoman Sultan Mastafa III, via his ambassador to Russia, Janer Pasha, gifted Catherine the Great an emerald powder box rubbed with smallpox scabs in an attempt to sicken, or even kill, their archrival. But the Tsarina was

Fig. 1.2 Eradication of Smallpox
© United Nations Postal Administration

apparently unscathed, possibly because she was already vaccinated. She continued to reign Russia wisely for nearly another thirty years without suffering from smallpox.

History is full of irony. A Russian heir, Tsar Peter III, loved Prussia and was inflicted with smallpox. Meanwhile his wife and usurper, a German princess, Catherine the Great, loved Russia and promoted smallpox inoculation. She "dragged" the twenty million subjects of backward Russia out their medieval stupor to join the enlightenment of Western Europe. She modernized Russia during her thirty-four-year reign. And the rest, as they say, is history.

1.1.4 The Louisiana Purchase

Infectious diseases were by no means only favored by European colonials during the clashes between the Old World and the New. Yellow fever, malaria, and other tropical diseases furnished the most formidable obstacles to European colonization. Yellow fever, so named because suffers' livers are damaged to cause jaundice, kills about ten percent of its patients due to liver failure. The mortality rate can be up to fifty percent. The yellow fever virus hampered Napoleon Bonaparte's New World dream and catalyzed a momentous event for American expansion: the Louisiana Purchase.

One of the most important events making the United States of America the country it is today was the Louisiana Purchase from France in 1803. The country nearly doubled its territory at the price of eighteen dollars per square mile. Financial, diplomatic, and military factors worked in concert to help President Thomas Jefferson succeed in completing the transaction. Infectious diseases, namely malaria and yellow fever, also played an important role to prompt Napoleon Bonaparte to sell the Louisiana Territory to the United States.

Yellow fever originated in Africa, where a number of primate species were infected. The yellow fever virus was then transported from West Africa to the West Indies during the sixteenth century, probably through the slave trade. Like malaria, yellow fever is transmitted by mosquitos. Unlike malaria, which is caused by protozoa, yellow fever is caused by a virus, *yellow fever virus*. Because mosquitos were the vector, yellow fever—also known as Yellow Jack—was rampant in tropical climates. The toll that yellow fever exacted on the French troops on the fateful Caribbean island of Saint-Domingue at the beginning of the nineteenth century shaped the Louisiana Purchase.

In 1802, on the American side, Jefferson was increasingly concerned with the control of the Mississippi River in general and the port city of New Orleans in particular. In 1762, the vast land of the Louisiana Territory was given as a gift to Charles III of Spain by Louis XV of France. After nearly forty years of Spanish administration, the First Consul Napoleon took it back in 1800 through the treaty of San Ildefonso because Napoleon's grande armée dominated Spain, and the whole of continental Europe for that matter, at the time. Confusions ensued during "the change of management," and transportation of American goods on the Mississippi and export from New Orleans to Europe looked in danger. While many Americans felt that they needed force to take over New Orleans from the French, the French troops were having troubles in Saint-Domingue.

Saint-Domingue (today's Haiti) was the western portion of the island called Hispaniola when Christopher Columbus "discovered" it in 1492, after which it subsequently became a Spanish colony. Because massacre, hunger, fear, and hard labor decimated the enslaved local Taino inhabitants, African slaves were brought to work in sugar and coffee plantations. Envious of the riches on the island, the French began to encroach upon the western portion of the island and Tortuga Island in the seventeenth century. Spain tacitly ceded the region to France in 1697 and it became Saint-Domingue, while the remainder of the island became Santo Domingo (today's Dominican Republic). Its sugar and coffee, produced at the cost of the freedom and lives of slaves, made the island a crown jewel of the

French West Indies. Its exports made the colonists wildly rich and filled France's national coffer. In 1798, a slave rebellion led by the charismatic Toussaint L'Ouverture liberated both Saint-Domingue and Santo Domingo and then fended off invasions by both the British and Spanish empires. The whole island became an independent nation.

Napoleon at first aspired to establish an empire in the Americas. In 1801, he sent General Charles Leclerc and thirty-three thousand veterans to occupy New Orleans. But they had to make a "detour" on Saint-Domingue to suppress Toussaint's slave uprising. Leclerc, Napoleon's brother-in-law, was married to Napoleon's frivolous sister, Pauline. He and his troops arrived at Saint-Domingue in the early 1802 and quickly had the situation under control by capturing Toussaint L'Ouverture and many of his generals. But the month of May brought rains and mosquitos, and along came malaria and yellow fever. While the black slaves from Africa had already developed a certain immunity against the yellow fever virus and they were hardly affected. The fresh troops from continental Europe, however, had not acquired the slightest immunity and the French soldiers died in droves. In fact, three thousand men died in a month, and after that the pace quickened. One regiment landed with 1,395 men and had only 190 left alive, with 107 of them hospitalized.[16] Almost all regiments had similar statistics. Leclerc was initially weakened by malaria and was finished off by the Yellow Jack in November 1802. A year later, when France withdrew its seven thousand surviving soldiers, more than two-thirds of the troops perished there. That was apparently not the first time that yellow fever decimated invading newcomers. Back in 1793, thirteen thousand British soldiers were nearly wiped out by the double whammy of malaria and yellow fever when they tried to take over Saint-Domingue from the French.

With the disaster in Saint-Domingue and the fear of a potential Anglo–American alliance, Napoleon decided to sell the Louisiana Territory for some quick cash. After all, his war with England was going to cost him a lot of money. Under the leadership of President

Jefferson and Secretary of State James Madison, American envoys James Monroe and Robert Livingston negotiated with Napoleon's foreign minister Charles Talleyrand and finance minister François Marbois, who happened to have served as the governor of Saint Domingue for five years during his youth. The Treaty of the Louisiana Purchase was signed in April 1803, thus concluding one of the most consequential real estate transactions for the United States, notwithstanding New York, which was bought for twenty-four dollars in 1626. The money from the Louisiana Purchase apparently helped Napoleon win the battle of Austerlitz, which was the pinnacle of his military career.

In 1898, by the Treaty of Paris, Spain renounced her claims to lands discovered by Columbus. Cuba ceased to be a Spanish colony and became a neo-colonial to the United States. However, through the sugar trade, Cuba brought yellow fever to the United State and caused more than nine yellow fever epidemics in the southern states, especially the states neighboring the Gulf of Mexico. A Cuban epidemiologist, Cárlos Finlay, made significant contributions to the etiology of yellow fever. Even today, Cubans believe that it was Finlay who, in 1881, discovered the mosquito as the vector of yellow fever. Major Walter Reed led the US Army's Yellow Fever Commission and eliminated bacteria and fomites as the infectious agents. Building on Finlay's knowledge, they firmly established in 1901 that the yellow fever virus was indeed transmitted by mosquitos. By removing the breeding grounds for these insects, the disease could be prevented. American microbiologist Max Theiler at the Rockefeller Institute developed a live avirulent strain of the yellow fever virus for use as a vaccine in the 1930s (see Figure 1.3). In 1951, Theiler received the Nobel Prize for "his discovery concerning yellow fever and how to combat it." Interestingly, yellow fever was the first human disease shown to be caused a *filterable agent* smaller than any known bacteria. A "filterable agent" was *virus* in today's term.

Fig. 1.3 Max Theiler
© Federated States of
Micronesia Postal Services

1.2 Discovery of Viruses

Bacteria were relatively easier to discover because they are bigger, whereas the discovery of viruses was harder because they are too small to be seen under normal optical microscopes.

1.2.1 The germ theory

We *homo sapiens* have always wondered what causes diseases, especially infectious diseases. Many theories had been put forward. In ancient times, diseases were thought to be God's punishment. Hippocrates suggested that humoral imbalance was the culprit. Before the emergence of the germ theory, the miasmatic theory for infectious diseases was popular. The first genuine scientific evidence was probably Antony van Leeuwenhoek's observation of bacteria. In 1683, Leeuwenhoek, a Dutch drape merchant and a minor city official, recorded that he witnessed "small living animals" using a microscope that he ground himself. German physician Jacob Henle was probably the first scientist to formulate the modern germ theory in 1840. Suspecting that germs might be responsible for causing childbed fever for new mothers, Hungarian physician Ignaz Semmelweis promoted disinfection in obstetrical clinics in the 1840s by washing his hands with a chlorinated lime solution. At the time, spontaneous generation theory claimed that microorganisms could arise without parents and life was created anew from inanimate matter. But others believed that every living thing originated from living things with similar characteristics, which was echoed by Rudolf Virchow's Latin dictum: "*Omnis cellula e cellula*" ("only from cells arise cells").

Also in the 1840s, Austrian scientist Theodore Schwann carried out a series of experiments trying to debunk the spontaneous generation theory. However, his results were hard to duplicate and not too many people were convinced. Eventually, it was Louis Pasteur who solidly established germ theory in the 1860s through a series of brilliantly designed experiments that were easily reproduced by his peers. Pasteur began his foray into germ theory and microbiology with the fermentation processes. From his microscopic observations during the fermentation processes of alcohol into acetic acid and sugar into lactic acid, Pasteur concluded that fermentation was a biological process catalyzed by yeasts. This was in contrast to the textbook dogma at the time. Eminent chemists, such as Jöns Jakob Berzelius in Sweden and Justus von Liebig in Germany believed that wine, beer, and vinegar were produced through an exclusively chemical process and that germs only presented during decomposition or fermentation. After only two years of experimentations, Pasteur resoundingly proved that fermentation was indeed a biological process. He even suggested, although without evidence, that the same principles might contribute to the understanding of infectious diseases.

One of his *tour de force* experiments in 1860 was the swan neck flask experiment. He placed fermentable fluid, such as meat broth, in a flask attached to an S-shaped tube. Heating the liquid would sterilize the swan neck and trap microbes at the tube. Without microbes, the fermentable broth would stay clear for a long time. In fact, 100 years after his experiments, some of Pasteur's flasks can still be seen at the Pasteur Institute in Paris, the fluid as limpid as the day it was sterilized. Moreover, Pasteur was also a consummate debater, and through his eloquence and his experimental results, he triumphed over famous German chemists Friedrich Wöhler and von Liebig and germ theory began to be widely accepted in the scientific community.[17]

Long before his death, Pasteur asked his family never to show his laboratory notebooks to anyone. A history professor at Princeton University, Gerald Geison, scrutinized Pasteur's laboratory notebooks and published a book on the subject in 1995.[18] From Professor Geison's systemic analysis, many discrepancies were found between Pasteur's experimental results and his formal scientific publications and demonstrations. Pasteur used strong rhetoric, exaggeration, and even fraudulent claims to buttress his germ theory and other topics.

One of Pasteur's contemporaries and archrival, Robert Koch, was a German doctor who founded bacteriology. Koch, a student of Jacob Henle, made his name with his discovery of the complete life cycle of *anthrax bacillus* in 1876. He went on to isolate *tubercle bacillus* in 1882 and *cholera bacillus* in 1883. He astutely summarized experimental observations concerning germ theory by delineating what was required for a specific germ to cause a specific disease. His criteria became so famous that they were soon dubbed Koch's four postulates:

- The bacterium must be present in every case of the disease;
- The bacterium must be isolated from the disease host and grown in pure culture;

- The specific disease must be reproduced when a pure culture of the bacterium is inoculated into a healthy, susceptible host; and
- The bacterium must be recoverable from experimentally infected hosts.

Koch's four postulates were warmly embraced by the scientific community in general and the bacteriologists in particular, which significantly accelerated the progress of bacteriology. Another reason why Koch's four postulates were popular was because people, including scientists, like things to be orderly: a set of postulates, rules, and theories just make life so much easier. The peril of such belief is when it becomes dogma. Dogma actually hinders the progress of science, as we will see in the discovery of viruses and many other scientific discoveries.

The germ theory gained almost universal acceptance in Pasteur's lifetime. Bacteria were much better understood because they could be "seen." But viruses were more challenging at the time, even though Pasteur had developed a vaccine for rabies caused by rabies virus. In 1884, working in Pasteur's laboratories, Charles Chamberland perfected the Chamberland filter, also known as the Chamberland filter-candle. It was an unglazed porcelain bar with pores smaller than bacteria so it could completely remove bacteria from a solution. For a long time, "filterable substances" were a synonym to viruses. The Chamberland filter contributed significantly to the discovery of viruses.

1.2.2 Discovery of viruses

The end of the nineteenth century was the heyday of virus discovery. Ironically, the discovery of viruses is attributed to the work of three botanists.

Adolf Mayer, a German scientist working in Holland, began his research on diseases of tobacco in 1879. Even though he did not discover the tobacco mosaic virus, he inoculated healthy plants with the juice extracted from grinding up leaves from diseased plants. By that, Mayer was the first person who demonstrated the first experimental transmission of a viral disease in plants. He coined the name *tobacco mosaic disease* and incorrectly suggested it was caused by bacteria. Koch's four postulates did not apply because the infectious substance here was neither bacteria nor fungi and the virus could not be isolated at the time.

In Russia in 1890, twenty-five-year-old microbiologist Dmitry Ivanovsky studied tobacco mosaic disease. He reported to the St. Petersburg Academy of Sciences: "I have found that the sap of leaves attacked by the mosaic disease remains its infectious qualities even after filtration through Chamberland filter-candle." The virus was later known as *tobacco mosaic virus*. This was the first evidence of the existence of viruses, although Ivanovsky did not grasp the significance of his observation, suspecting that the filters he used must have had fine cracks that allowed small microbial spores to pass through. Nevertheless, Ivanovsky is known today as the father of virology. Six years later, in 1898, Adolf Mayer's young protégé, Martinus Beijerinck, in Holland, independently repeated Ivanovsky's experiment and confirmed that the filtered infectious substance was not a toxin. Beijerinck believed that the causative agent was an infectious liquid and christened the filtered infectious substance with the term "virus," which means "slimy liquid" or "poison" in Latin.[19] It was also in 1898 that the passage of an animal pathogen through a Chamberland filter yielded the foot-and-mouth disease virus (an RNA virus) of cattle, the first animal virus. It was discovered by German bacteriologist Friedrich Loeffler, a student of Robert Koch (see Figure 1.4). Loeffler correctly concluded that the agent causing the foot-and-mouth disease was a small particle that passed through his Chamberland ultrafilter.

Fig. 1.4   Friedrich-Loeffler © Deutsche Post

Many believe that Loeffler was the true founder of virology. He was also the one who popularized guinea pigs as an animal model for medical science in 1884, when he discovered these adorable little Andean creatures were highly susceptible to germs.

The story of the tobacco mosaic virus did not end there. In 1946, biochemist Wendell Stanley, at the Rockefeller Institute of Medical Research in Princeton, won the Nobel Prize for his achievement of the crystallization of the tobacco mosaic virus. Back in 1935, Stanley purified the tobacco mosaic virus and derived fine, needle-like crystalline preparations that were fully active and infectious. Some viruses can be crystalline thanks to their highly symmetrical structures. Not only was it the first time a virus was crystalized, but it was also an unprecedented and unexpected example of a living entity occurring in the crystalline state. In 1956, Heinz Fraenkel-Conrat in the United States and Alfred Gierer and Gerhard Schramm in Germany showed independently that the tobacco mosaic virus was a single-stranded RNA virus of filamentous morphology, and that RNA alone was sufficient to allow reproduction of the virus in infected leaves. In 1966, tobacco mosaic virus's rod-like crystal structure was visually observed using electron microscope.

It did not take long for scientists to determine that viruses are the causative agents for many infectious diseases. Smallpox, rabies, polio, yellow fever, chickenpox, shingles, influenza, and common colds are all caused by viruses. Later, we discovered the pathogens causing hepatitis, acquired immunodeficiency syndrome (AIDS), Ebola, and severe acute respiratory syndrome (SARS). COVID-19 is also caused by a virus, the severe acute respiratory syndrome-coronavirus-2 (SARS-Cov-2).

For a long time, it was widely believed that the RNA virus was the smallest "live" agent—until the discovery of the *viroid*. While working on potato spindle tuber disease, plant virologist Theodor Diener, at the United States Department of Agriculture (USDA), discovered particles that were one-fiftieth of the smallest viruses.[20] Viroid, a rare occurrence, is not to be confused with the ubiquitous *virion*, which is a complete virus before entering the host cell to infect. All viruses have virions, which have either RNA or DNA molecules wrapped around by a capsid, which is the protein envelope.

1.2.3 Cancer-causing viruses

Seven human viruses have been found to cause fifteen to twenty percent of human cancers worldwide. But in the early 1900s, human tumor virology was unheard of. The concept of a virus causing cancer was often disregarded, if not marginalized and ridiculed. Skepticism was prevalent. Even the carcinogen theory for cancer's etiology did not start to gain credence until Katsusaburo Yamagiwa's famous experiment in 1915. He applied coal tar on rabbit ears and had observed tumor formation after several months. At the time, the prevalent view of cancer etiology was the *intrinsic hypothesis*, where cancerous growth was due to the disruption of the function of the inner mechanism of the cell.

In 1908, two Danish scientists, Vilhelm Ellerman and Oluf Bang, described a form of leukemia in chickens that could be passed as a filterable agent. Remember that because viruses had not yet been defined, "a filterable agent" referred to a virus. That was the first example of an avian leukemia virus. But because the malignant nature of leukemia was not recognized until decades later, Ellerman and Bang's discovery was generally not recognized either. It was also perceived to be irrelevant to humans because chickens are not even mammals. One year later, another discovery made by an American experimental pathologist, Peyton Rous, had better luck and garnered a little bit more attention from the scientific community. *A little bit* is the operating phrase here.

In 1909, Rous was a young MD working at the Rockefeller Institute as a researcher. A lady from the neighboring farm in Long Island brought in a fifteen-month-old barred Plymouth Rock hen. The chicken had a large muscle sarcoma (a sarcoma is a tumor of soft tissue, such as a muscle or nerve) in her right breast. Using the then-popular tumor transplantation technique, Rous succeeded in transplanting the tumor to other hens in the same bloodline by injecting small pieces of the growth. To carry out his research on this topic, Rous bought many more chickens from the same chicken breeder by offering a reward of one dollar apiece. Later, he achieved the same feat as part of his "scientific duty" by injecting cell-free filtrates (virus) that passed through a Berkefeld filter, which is a very fine porcelain filter that held back miniscule particles, such as bacteria or tiny fragments. The filterable agent would not pass through the fine filters, such as a dialyzing membrane and the Chamberland bougies that were much finer than Berkefield filter and therefore retained microorganisms in their walls. This was a good indication that the filterable agent was of a living nature. Rous and his associates subsequently identified about sixty more naturally occurring "true" tumors. The tumor virus was later named the Rous sarcoma virus. After four long, fruitless, and frustrating years, Rous was unsuccessful in isolating the virus, which compelled him to discontinue his work on cancer viruses. It was not until fifty-five years later, at the age of eighty-nine (and still working!), that Rous was bestowed the 1966 Nobel Prize in Medicine for "his discovery of tumor inducing viruses." He is widely considered the father of tumor virology.[21] His fellow Nobel laureate that year was Charles B. Huggins, of the University of Chicago, for his discoveries concerning the hormonal treatment of prostatic cancer. Today we know that Rous was also lucky with the Rous sarcoma virus because the virus carried at least two sets of genes, one for replicating itself and the second for inducing cancer. In any event, the virus pulled a double duty! The Rous sarcoma virus can only cause cancer in tangent with an oncogene that it carries by itself as a payload. In 1965, Peter Duesberg and William Robinson at Berkeley were the first to isolate the relatively large-sized RNA intact from the Rous sarcoma virus.

After Rous' publication in 1911, many mammalian tumor-inducing viruses were found in mice, cats, rabbits, and monkeys throughout the next few decades. Nonetheless, attention to these discoveries was also still lacking because they were just animals. Unreceptive scientists would say: "Everyone knows that cancer is not an infectious disease. Wake me up when you find a tumor-inducing virus that causes cancer in humans!" That happened in 1964 when the Epstein-Barr virus, a DNA tumor virus, was discovered.

In 1958, pathologist Anthony Epstein at Middlesex Hospital Medical School in London demonstrated the Rous sarcoma virus was an RNA, not a DNA virus, using the newly available electronic microscope to study its morphology. On March 22, 1961, Epstein was enthralled by a seminar given by Denis Burkitt, an Irish surgeon from Mulago Hospital in Kampala, the capital of Uganda. Burkitt described a strange malignant tumor in children that was affecting bizarre sites, mostly in jaws and abdomens, that was fatal within a few months. Burkitt achieved spectacular results using chemotherapies, including methotrexate, cyclophosphamide, and vincristine. To which Burkitt claimed: "the most ignorant chemotherapist in the world getting the best results."[22] The cancer was later coined as Burkitt's lymphoma. Lymphomas are tumors that arise from white blood cells, known as lymphocytes (natural killer cells, T-cells, and B-cells), in our immune system. Because this disease was a cancer of the B-cells that normally produce antibodies, Burkitt's lymphoma patients were found by closely following the African malarial belt.

Epstein immediately formulated his hypothesis, even before Burkitt's seminar was over. He correctly suspected that it was a climate-dependent arthropod vector spreading a cancer-causing virus. Epstein spoke with Burkitt after his seminar, and they began a collaboration to find the virus that caused Burkitt's lymphoma.

Receiving cancer samples from Burkitt in Kampala, Epstein attempted to identify and rescue the putative virus living within Burkitt lymphoma cells. He found no virus in three years, but one chance encounter changed their fortune. In the summer of 1963, a grant from the United States' National Institute of Health (NIH) enabled Epstein to hire two associates—Miss Yvonne Barr and Mr. Bert Achong. In December 1963, a fog-delayed flight from Kampala delivered a Burkitt lymphoma biopsy later than had been expected. The transit medium in which the biopsy was shipped was unusually cloudy. Rather than suspecting contamination of bacterial contamination, Epstein examined the sample using a light microscope and observed a large numbers of viable tumor cells floating free of the main lymphoma mass. The tumor cells had been shaken off from the cut edge of the lymphoma sample during the flight. These cells grew while suspended in the fresh culture medium over twenty-six days, allowing the culture to be split into new culture bottles. The first cell line from human lymphoma was established. Using an electronic microscope, Epstein observed in the very first grid square a cell filled with herpesvirus. Considering it was later found that only one percent of the Burkitt lymphoma cells had the virus particles, Epstein was really lucky that day. Thus, the first human tumor virus was discovered.[23] Immediately upon realizing the momentous discovery that he had made at that instant, Epstein reminisced: ". . . I was so amazed and euphoric that I was terrified that the specimen would burn up in electron beam. So, I switched off, I walked round the block in the snow without a coat, and somewhat calmer I returned to record what I had seen." He wrote down that the electronic microscope had also revealed that the viruses were "virus, like herpes," thus establishing the virus as a new human herpesvirus.[24] The event was emblematic of Pasture's sage words: "In the field of observations, chances favor prepared minds."

When Epstein submitted the paper for publication, expert referees were unwilling to believe that human lymphocytic cells could be cultured at all because it was the first time such a feat was accomplished. Of course, suspension is now the standard technique to grow such cells and is used worldwide for a huge number of different types of research. The virus was a DNA tumor virus, which came as a surprise to most scientists in the field of viral oncology because all cancer-causing viruses in animals were RNA viruses. Epstein's virus was later coined as the Epstein-Barr virus because the cell sample was labeled as Epstein-Barr cells, as both Epstein and Barr were working on it. Even though another assistant, Achong, was a coauthor of the original publication along with Epstein and Barr, was left out as part of the virus name. That paper became one of the most cited papers in scientific publications.[25] Interestingly, Epstein-Barr virus was the end of the tradition of naming viruses after their discoverers.

Epstein-Barr virus was the first human virus ever discovered by electronic microscope and it was also the first human cancer virus. Later, Werner and Gertrude Henle at the Children's Hospital in Philadelphia identified Epstein-Barr virus as a ubiquitous virus infection (most of us carry the virus, which harmlessly coexists with us), and as the cause of infectious mononucleosis (mono, or the kissing disease), and as a transforming virus. Burkitt lymphoma, on the other hand, was one of the first human tumors to be linked with a specific set of chromosome translocations.

Another cancer-causing virus is the hepatitis B virus. In 1968, Baruch Blumberg and colleagues demonstrated that blood from hepatitis patients contained the Australian antigen, an antigen first isolated from a blood sample from a member of the Australian indigenous people. His seminal work proved that the Australian antigen was the surface antigen of a *hepadnavirus* to be called hepatitis B virus, the virus that caused hepatitis B. Hepatitis B virus was found to work in tangent with another risk factor to cause hepatocellular carcinoma. Blumberg was awarded the Nobel Prize in 1976 for his landmark discovery of the Australian antigen.[26] In 1988, the hepatitis C virus was discovered; it has also been linked to hepatocellular carcinoma.

Human papillomavirus (HPV) is a DNA virus that causes approximately five percent of human cancers, most conspicuously cervical cancer in women. In 1907, Italian physician Giuseppe Ciuffu reported that a cell-free filtrate derived from benign papilloma (common human warts) could transfer disease. The papillomavirus was initially studied in animals. In 1935, Peyton Rous and his young colleague at Rockefeller Institute, Joseph Beard, showed that the cottontail rabbit papillomavirus caused skin carcinoma in rabbits.

Among all human malignancies, cervical cancer (initially called genital warts) is the most likely to present as a sexually transmissible disease. Harald zur Hausen, a doctor in Heidelberg, Germany, discovered that HPV caused cervical cancer in women. In 1974, based on the observation that genital warts infrequently become malignant, zur Hausen proposed that HPV could represent the etiologic agent for cervical cancer. In the early 2000s, zur Hausen found novel HPV DNA in cervical cancer biopsies and he subsequently cloned HPV16 and HPV18 subtypes from patients with the disease. HPV16 and HPV18 account for two-thirds of all cervical cancers worldwide. Cervical cancer is the second most common cancer among women, behind breast cancer. In 2008, zur Hausen

Fig. 1.5 Harald zur Hausen © Union of
Comores Post Office

was bestowed the Nobel Prize for his work on HPV (see Figure 1.5).
By 2009, Merck's HPV vaccine, Gardasil, was approved by the FDA.
Gardasil is a recombinant human quadrivalent papillomavirus
vaccine (types 6, 11, 16, and 18). In the same year, the FDA also
approved GlaxoSmithKline's Cervarix, which is a di-valent HPV
vaccine (types 16 and 18).

    After all was said and done, the virus theory for cancer eti-
ology fell short because of its failure to explain the vast majority
of human cancers which have no viral association at all. Thanks to
the work of numerous scientists in virus and cancer research with
significant contributions from virologists, we have now amassed
enough knowledge to better appreciate the etiology of cancer.
Although in some cases cancers may be caused by carcinogens
or viruses, cancer is ultimately a disease of genes. The concept
of proto-oncogenes, oncogenes, and tumor-suppressor genes
helps to convincingly explain the etiology of most cancers. In
1989, J. Michael Bishop and Harold Varmus won the Nobel Prize
for their discovery of the cellular origin of retroviral oncogenes,

Fig. 1.6 Harold Varmus and Michael Bishop © Republic of Palau Post Office

showing that normal genes in every cell can cause cancer if they are disrupted (see Figure 1.6).

1.2.4 Discovery of retroviruses

In addition to DNA and RNA viruses, there are also retroviruses. To be sure, a retrovirus is still an RNA virus because its genetic information is carried by the RNA molecule. Nevertheless, its genetic information flows from RNA to DNA before assuming the natural flow again to RNA and then to protein. Because the resulting DNA made from RNA is complementary to RNA in terms of nucleotide sequence, we call it *complementary DNA* (cDNA). All aforementioned RNA tumor viruses are retroviruses. For instance, Rous sarcoma virus is a retrovirus.

DNA viruses encompass hepatitis, Epstein-Bar virus, herpes virus, smallpox virus, and papilloma virus. RNA viruses include poliovirus, influenza virus, yellow fever virus, and a dozen other less common viruses.

"Virus is a piece of nucleic acid surrounded by bad news" according to Peter Medawar, the 1960 Nobel Laureate for Physiology or Medicines. Indeed, a virus uses either DNA or RNA molecules as the genome to archive the genetic information. And it uses a glycoprotein envelope which first attaches the virus itself to its host cell before entering. When it infects the next cell, the DNA or RNA molecules serve as a blueprint so that all its genetic information is passed on to the next generation. On a molecular level, it is well known how genetic information is stored and transmitted. The information is passed on from DNA to DNA or RNA via *transcription* and from RNA to proteins via *translation*.

In the late 1950s, Francis Crick, the co-discoverer of the double helix structure, advanced the *central dogma* for the flow of genetic information. Central dogma describes the flow of genetic information in a two-step process of transcription and translation in a unidirectional vector: DNA → mRNA → Protein. mRNA stands for messenger ribonucleic acid (see Figure 1.7).

Here comes a question. How does a retrovirus duplicate itself? The easy answer would be the RNA makes another RNA in the newly infected cell. But the reality is not that straightforward.

In the 1960s, Howard Temin at the University of Wisconsin proposed a "DNA provirus hypothesis." In contrast to the *central*

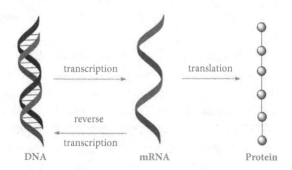

Fig. 1.7 The Central Dogma

dogma, he believed that the RNA tumor viruses, such as Rous sarcoma virus, reproduced themselves by going through a DNA form. In other words, an RNA molecule made a DNA copy of its genes first after it invaded a host cell. The RNA molecule was used as a template for constructing the new DNA molecule. After all, the RNA of the Rous sarcoma virus is not infectious. Moreover, DNA is more robust than RNA, therefore DNA is a more reliable vehicle for carrying the genetic information.

Schematically, Temin proposed that the genetic information for retroviruses flows like:

RNA → DNA → mRNA → Protein

But Temin's theory could not convince anyone else without robust experimental evidence. After all, no DNA could be detected in RNA viral particles. In 1969, Temin's group carried out an interesting experiment using actinomycin D, a metabolic poison that would attack DNA but would leave RNA untouched. It worked by intercalating into double-stranded DNA-directed, but not RNA-directed, RNA synthesis thus it could block the production of virus RNA by previously infected cells. After treating cells infected by Rous virus with actinomycin D, they observed that actinomycin D blocked the growth of Rous virus. This experiment indirectly confirmed the presence of viral DNA inside infected cells. But the presence of viral DNA, the product of the copying reaction, was weak evidence.

In Temin's lab at the time, a Japanese postdoctoral fellow, Satoshi Mizutani, although not good at cell and virus culture, was an excellent biochemist. Once they knew what they were looking for, the experiments were easy to carry out, as they only required a detergent, some buffer solutions, and some radiolabeled deoxynucleotide triphosphates. They also realized that the enzyme would only appear after the virions entered the host cells. Mizutani succeeded in detecting the enzyme that converted RNA to DNA: a polymerase that they initially christened as an RNA-dependent DNA polymerase; the name was soon changed to reverse transcriptase. It was there all along in the virus particles, waiting to be found! The experiment was extremely trivial to do—it would only take a few hours—if you knew where to look. Almost simultaneously, David Baltimore at MIT, working on poliovirus, also found the viral copying enzyme polymerase in the virus particles but he called it a replicase. It was later merged as reverse transcriptase.

Real life events are often more dramatic than fiction. Baltimore's manuscript arrived at the prestigious British journal *Nature* on June 2, 1970, and Temin's arrived thirteen days later, on June 15. The journal published their papers back-to-back on the June 27 issue. The speed with which their papers was published spoke volumes about the importance of their work. Temin and Baltimore won the Nobel Prize in 1975, a short five years after their discovery.[27] Their discovery revolutionized molecular biology and laid the foundations for retrovirology and cancer biology. Gratifyingly, reverse transcriptase played a decisive role in the discovery of human immunodeficiency virus (HIV) in 1982. This is also a good example of basic scientific research having a profound impact on real life.

The first human retrovirus, human T-cell leukemia virus (HTLV), was discovered by Robert C. Gallo's group at the National Cancer Institute (NCI), a division of NIH. Gallo's student Bernie Poiesz isolated and identified HTLV from the lymphocytes (white blood cells) of a patient with cutaneous T-cell lymphoma.[28] Independently, Japanese scientists also discovered HTLV from their studies of adult T-cell leukemia. A small population of residents in Kyushu (九州) Island on the southwestern tip of Japan had higher rates of a leukemia, a rare and exotic cancer. Yorio Hinuma, Mitsusaki Yoshida, and Isao Miyoshi isolated and identified HTLV in 1982. It was later established that HTLV was the etiologic agent for adult T-cell leukemia.[29] HTLV is arguably one of the most carcinogenic agents.

Retroviruses include Rous sarcoma virus, infectious equine anemia virus, HTLV, simian immunodeficiency virus (SIV), and human immunodeficiency virus (HIV), which happens to be a topic of Chapter 2.

2

HIV/AIDS Drugs

Transforming a Death Sentence into a Chronic Disease

All things are hidden, obscure and debatable if the cause of the phenomena be unknown, but everything is clear if this cause be known.

—*Louis Pasteur (1822–1895)*

HIV and AIDS are such household vernacular terms that they need no further explanation. However, in case the reader is too young, let's do a quick review. HIV stands for *h*uman *i*mmunodeficiency *v*irus. HIV is the pathogen that causes AIDS, or *a*cquired *i*mmunodeficiency *s*yndrome. An AIDS patient's immune system breaks down, leaving the victim vulnerable to opportunistic infections by bacteria and fungi that would not normally be lethal to humans. It is believed that AIDS crossed species from a chimpanzee in the 1950s in Central Africa. The disease then spread from Africa via Haiti to the United States in the 1970s.

Despite the tremendous strides that we have made in antiretroviral drugs, AIDS remains a scourge for humanity worldwide today. AIDS patients in poor countries still have difficulties accessing life-saving medicines. Since the beginning of the HIV/AIDS pandemic in 1981, AIDS is estimated to have killed more than twenty-five million people. According to the United Nations' statistics,

Conquest of Invisible Enemies. Jie Jack Li, Oxford University Press. © Oxford University Press 2022.
DOI: 10.1093/oso/9780197609859.003.0002

approximately forty million people were living with HIV in 2019.[1] Notably, AIDS has replaced both malaria and tuberculosis as the world's deadliest infectious disease. Even today, the United States sees about forty thousand new infections annually.

As far as HIV is concerned, it is a *retrovirus* and its genetic material is *ribonucleic acid* (RNA). Once the virus enters its host cells, its genetic information flows from RNA to *deoxyribonucleic acid* (DNA) before resuming the natural flow, according to Francis Crick's *central dogma*. The presence of the reverse transcriptase is unique to retroviruses and indeed was vital to the discovery of retroviruses, especially HIV.

2.1 Discovery of the Virus

With the exception of a tiny fraction of people on the fringes, it is now nearly universally accepted that HIV is the virus that causes AIDS infection. But in the early, mystifying days, what exactly caused AIDS was hotly debated. Some initial ideas put forward were utterly absurd!

2.1.1 A death sentence

The first AIDS patient in the United States was recorded in 1981 at the University of California at the UCLA hospital in Los Angeles, although the term AIDS did not exist yet at the time. Dr. Michael Gottlieb, a clinician, saw a patient inflicted with pneumocystis, a very rare and severe form of pneumonia. He reported the disease as "acquired T-cell defect" because the patient had no T-cells left in his system. It was as though his entire immune system had been wiped out. T-cells and B-cells are two major cellular components of human *adaptive immune response*. T stands for *thymus* because

T-cells were at first isolated from thymus. B stands for *bursa of Fabricius* in birds. B-cells are a type of white blood cell that are responsible for killing invading pathogens, such as bacteria and viruses.

Soon after Gottlieb's report, more and more similar cases of pneumocystis began to pop up among homosexual men in San Francisco and New York, along with Kaposi's sarcoma, which is a very rare and severe form of cancer. The disease was at first called *gay-related immunodeficiency* (GRID); the name was changed to AIDS in 1984. The Center for Disease Control and Prevention (CDC) in the United States estimated the fatality rate of AIDS at that time to be nearly one hundred percent. Having contracted AIDS was equivalent to a death sentence.

But what exactly caused the infectious and deadly disease?

Some scientists, including Peter Duesberg at the University of California at Berkeley, suggested non-infectious causes. Duesberg made a name for himself when he isolated the relatively large-sized RNA intact from the Rous sarcoma virus. It was speculated that AIDS was possibly caused by amyl nitrate (poppers), a stimulant and an aphrodisiac popular with gay men to enhance sexual pleasure. Some others proposed that AIDS was caused by auto-immune reactions against foreign white blood cells transmitted by rough sexual practice. It seems ridiculous now, but some researchers conjured that sperm was the cause of the disease because semen was known to be immune-suppressing. Even a defective thymus gland was proposed as the possible cause of AIDS because of the T-cell defects were associated with the disease. The etiology of this particular disease became more and more puzzling. At one point, a fungus that released cyclosporine was contemplated as a potential root of the disease. After all, cyclosporine is a T-cell inhibitor.

Later, similar cases started to appear in hemophiliacs and blood transfusion patients. That observation immediately ruled out any theory that the disease was solely sexually transmitted because it was now clear that blood transfusion was another viable way of transmitting the disease. Hemophiliacs receive transfusions of the clotting factor VIII and its production goes through several filtrations. Big particles, such as bacteria, fungi, and protozoans, would be removed. Therefore, only a smaller organism—a virus—could be the causative agent. Many viral infectious theories were proposed as the cause of AIDS. Adenovirus isolated from the urines of some AIDS patients, Epstein–Barr virus, cytomegalovirus, and herpes virus were all suspects at some point.

It became widely accepted that HIV was the causative agent of AIDS in 1984 after the virus was discovered. In 1987, AZT (azidothymidine, zidovudine, Retrovir) emerged as the first antiretroviral drug, and now six classes and dozens of drugs are available for HIV/AIDS treatment. In 1996, highly active antiretroviral therapy (HAART), also known as cocktail drugs, transformed AIDS from a death sentence to a chronic disease that can be managed. This treatment comes close to a cure as measured by sustained viral response (SVR). David Da-Yi Ho (何大一), who pioneered the HIV cocktail regimen, was chosen to be "Person of the Year" in 1996 by *Time* magazine.

2.1.2 Discovery of HIV and the great scientific controversy

The discovery of HIV was important by itself for identifying the causative pathogen of AIDS, which was a prerequisite for developing blood tests to screen for the HIV viral infection and for discovering effective treatments. Those who discovered the virus no doubt deserve our recognition and appreciation. Luc Montagnier and Françoise Barré-Sinoussi at L'Institut Pasteur in Paris were the first to discover HIV in 1983. Meanwhile, Robert C. Gallo at the National Cancer Institute (NCI) also played an important and dramatic role during the discovery. Real life is often more dramatic than fiction. Together, let us trace back how the saga unfolded almost four decades ago.

2.1.2.1 Robert C. Gallo

In 1965, Robert C. Gallo joined the NCI, a division of the National Institute of Health (NIH). In a few years, he quickly rose to be chief of the Laboratory of Tumor Cell Biology. His goals were initially to discover and investigate cancer-causing viruses. He soon changed the direction for his research in 1970 when Howard Temin and David Baltimore's discovery of reverse transcriptase revolutionized molecular biology and laid the foundations for retrovirology and cancer biology. Gallo, like many of his peers, joined the fray by jumping into the field of retroviruses after listening to an electrifying seminar given by Baltimore at the NIH.

In the spring of 1976, Gallo announced that his group had discovered a human tumor virus. But he quickly fell on his face while presenting his results at the annual Virus Cancer Program in Hersey, Pennsylvania. Scientists who received his virus samples for independent confirmation revealed, one by one, that his "human" virus was actually a contamination of animal viruses. One of his cell lines was contaminated by three different animal primate retroviruses: woolly monkey virus, baboon endogenous virus, and gibbon ape leukemia virus. What a fiasco! Humiliated, Gallo picked himself up and forged ahead. A year later, he and his two postdoctoral fellows, Doris Morgan and Frank Ruscetti, discovered a T-cell growth factor, which was later renamed as interleukin-2. It is important to understand that growth factors are of critical importance when growing viruses, among many other biological functions. Rita Levi-Montalcini and Stanley Cohen received the 1986 Nobel Prize for their discovery of epidermal growth factor from epidermal cells of skin. The epidermal growth factor receptor (EGFR) inhibitors such as AstraZeneca's gefitinib (Iressa), Genentech's erlotinib (Tarceva), and AstraZeneca's osimertinib

(Tagrisso) are all tyrosine kinase inhibitors prescribed as targeted cancer treatments.

Then, 1980 was a good year for Gallo. His postdoctoral fellow, Bernie Poiesz, isolated and identified the first human retrovirus, a human T-cell leukemia virus (HTLV), from lymphocytes (white blood cells) of a patient with cutaneous T-cell lymphoma (CTCL). Lymphoma is cancer of the lymphatic system, including the lymph nodes, spleen, thymus gland, and bone marrow. Independently, Japanese scientists also discovered the HTLV during their studies of adult T-cell leukemia endemic in Kyushu (九州), Japan. Yorio Hinuma isolated and identified HTLV in 1981, confirming Gallo's discovery. It was later established that HTLV was the etiologic agent for adult T-cell leukemia. The discovery of HTLV was dependent on the availability of interleukin-2, Gallo's earlier discovery. Interestingly, HTLV's morphology, as observed by electron micrography, showed a large, spherical core typical of the type C retrovirus. This morphology was later used to distinguish HIV as different from HTLV because their morphologies are distinctively different.

In 1983, as the number of AIDS patients started to surge in the United States, the NCI director, Vincent DeVita, asked Gallo to ratchet up his research in the AIDS area. If all one has is a hammer, everything looks like a nail. Having discovered not one, but two human T-cell leukemia viruses—HTLV-1 and HTLV-2, both human retroviruses—Gallo was determined to contend that AIDS was caused by another variant of HTLV. Due to his influence and his alliance with the influential Myron "Max" Essex at Harvard, Gallo dragged the whole field with him down the wrong path. Both the NCI and the World Health Organization (WHO) believed at that time, incorrectly, that HTLV represented the best candidate for the etiology of AIDS. In retrospect, it was impossible for HTLV to be the etiologic agent of AIDS. First, hemophiliacs were contracting the virus from infusions of contaminated blood-clotting factors that were free of cells, and HTLV could not survive without T-cells. Second, AIDS patients did not always have T-cell leukemia either.[2]

2.1.2.2 Luc Montagnier and Françoise Barré-Sinoussi

In Paris, Luc Montagnier (see Figure 2.1) was already a veteran virologist when he and Françoise Barré-Sinoussi (see Figure 2.2) discovered the HIV in 1983.

At a height of five-feet-eight-inches, Montagnier (which means mountain in French) blamed his short stature on food deprivation during his youth in WWII. In his early career, he studied the foot-and-mouth disease virus and the Rous sarcoma virus, from which he discovered a double-stranded RNA, even though most RNA molecules are single-stranded. After joining the Pasteur Institute in 1972, Montagnier investigated interferon and carcinogenic retroviruses in humans. After Gallo's discovery of HTLV in 1980, Montagnier and Gallo initiated a collaboration to take advantage of Gallo's T-cell growth factor, interleukin-2.

By 1982, evidence began to emerge to suggest that the causative agent of AIDS was a retrovirus. Which one? Nobody knew! Because France imported some blood from America for their production of a hepatitis B vaccine, contamination by this retrovirus became a clear and present danger. At the end of 1982, Montagnier received a sample of blood and a biopsy sample of a lymph gland from a patient

Fig. 2.1 Luc Montagnier © Bhutan Post

Fig. 2.2 Françoise Barré-Sinoussi
© Union of Comores Post Office

with persistent lymphadenopathy. It was already known that AIDS destroyed T_4 lymphocytes (a special type of T-cell mostly affected by AIDS) and was preceded by lymphadenopathy—the swelling of some lymph nodes. T_4 cells are white blood cells of the lymphocyte lineage. Using the only one primitive wooden flow hood available at his disposal, Montagnier dissociated the biopsy tissue into single cells. To nourish the cell culture, he added interleukin-2, followed by a mitogen called phytohemagglutinin to promote cell division. For good measure, he added some anti-interferon serum, which could also stimulate virus growth. The addition of anti-interferon serum was not necessary and it was not used by other researchers to grow HIV. Every three days, his colleague, Dr. Françoise Barré-Sinoussi, would take an aliquot from the cell culture to detect the presence of retrovirus by measuring the reverse transcriptase enzyme activities. Nothing happened until the fifteenth day, at which point they began to see enzymatic activities of the reverse transcriptase in the lymph node cells, although not in the blood cells. Even though the cells began to die, Montagnier and Barré-Sinoussi were able to revive the cell culture by adding lymphocytes isolated from the fresh blood of a Spanish donor. In terms of the identity of this retrovirus in their hands, Gallo's HTLV was obviously a strong contender. After receiving Gallo's HTLV-specific antibody, it did not take long for Barré-Sinoussi and Montagnier to determine that their virus was not HTLV; they saw no cross reaction with each other. This was the first proof that the French had a different virus, which, although it was like HTLV, it was also an RNA virus. What they had was a brand new human retrovirus, even though the new virus had a density of 1.16 in a sucrose gradient, exactly the same as that of HTLV. Later, Montagnier recruited Françoise Brun-Vézinet and her assistant, Christine Rouzioux, to develop an enzyme-linked immunosorbent assay (ELISA) that used an antibody to detect the ligand (in this case, the virus) binding to the antibody. The ELISA assay, a serological test, was robust and greatly aided in the detection of HIV. The Pasteur team filed a patent for their blood test, which would become a source of acrimonious contention and which would lead to a three-year-long patent feud with Gallo.

At that time, an electron micrograph clearly showed that the virus had cylindrical or conical cores. That picture would, in time, become the icon of HIV, showing up on posters and postage stamps and in education materials and articles. Seeing the electron microscope image of the virus, a virologist colleague at Pasteur commented to Montagnier that it looked like an infectious equine anemia virus, which is a lentivirus or lentiretrovirus. *Lenti* means "slow" in Latin, and lentiviruses are a special class of very slow-acting retroviruses that cause autoimmune diseases—mostly chronic degenerative and neurological diseases—in horses, cows, goats, and sheep. The primate lentiretroviruses include the HIVs and the simian immunodeficiency viruses (SIVs). HIV infects humans whereas SIV infects African monkeys. Excited about the virus being a lentiretrovirus, Montagnier asked a veterinarian colleague for a sample of antibodies, which cross-reacted with the core protein of the virus. Therefore, the virus was indeed a lentiretrovirus, unrelated to HTLV. Because HIV is a lentivirus, it may take ten to fifteen years for some HIV-positive patients to develop symptoms, while others quickly develop AIDS soon after infection. Furthermore, the Pasteur team also found that the virus had a selective tropism to CD4-positive T-cells. This meant the virus infected CD4 + T-lymphocytes, the very cell affected in AIDS. CD stands for *c*luster of *d*ifferentiation.

Slowly, but surely, Montagnier and his colleagues at the Pasteur Institute accumulated enough data to ascertain that their virus was indeed a novel human retrovirus. They christened it as lymphadenopathy-associated virus (LAV), although this would later change to HIV. Normally, this would constitute a swan song of an extraordinary discovery; however, this was just the beginning of a great scientific controversy.[3]

2.1.2.3 Controversy

Although Montagnier had the *bona fide* virus, he and his colleagues met with a collective, resounding apathy from the scientific community at large. They had difficulties publishing their results in prestigious journals and had an even harder time convincing the world of the import of their discovery. Their own colleagues at the Pasteur Institute doubted that their virus was the cause of AIDS because Gallo and other world leaders in the field were barking at the wrong tree: HTLV. In reality, Gallo and Essex were only able to find HTLV in a small percentage of AIDS patients who were infected by both HIV and HTLV. One of the few exceptions to this lack of acceptance was the CDC's reception of Pasteur's virus. The CDC researchers, including Donald Francis and James Curran, were among the early believers of the French researchers and they actively collaborated with them.

In September 1984, Montagnier presented their discovery at a Cold Spring Harbor meeting on Long Island, New York; he received a cold reception. Few in the audience were convinced that they had the genuine virus that caused AIDS. It did not help at all that Gallo, as the meeting chair, had viciously challenged the validity of their discovery. (He half-heartedly apologized to Montagnier privately after the meeting.) Later, Gallo struck a more conciliatory tone and exchanged virus samples with Montagnier privately. In accordance with the practice common in the research community, the French group shared their work with many laboratories around world, including with Gallo's lab. In contrast, Gallo himself initially refused to share his viruses with anyone. When he did, he only gave it to people he liked and to those who would list him as a co-author. This was one of the reasons why Gallo was one of the most cited authors in the world in the 1980s.

While the French had success growing the virus using B-cells immobilized by the Epstein–Barr virus, Gallo's associate from Czechoslovakia, Mika Popovic, was not able to grow Montagnier's virus in T_4 tumor cell lines. He even had difficulties growing HTLV in cells. Probably frustrated by his inability to grow HTLV in cells, Popovic resorted to a dubious and highly controversial method of "pooling" many isolates from different patients. Apparently, he combined many of his samples together with Montagnier's sample. Accidentally or intentionally? We would never know. When an infectious agent reproduced faster than others, even if it was initially presented in a small quantity, it could quickly replace other variants. Lo and behold, Popovic and his lab succeeded in growing "their own" virus, which Gallo promptly christened HLTV-IIIB. In respect, Popovic used Montagnier's virus to infect a cell line from another source to carry out his Eureka! experiment. There was nothing wrong up to that point about what they had done. But claiming both the virus and the cell line to be their own new discoveries, as Gallo promptly did, constituted a double scientific fraud. Eager to bask in the glory of his new discovery, Gallo and the US Government announced to the world that he finally found the causative agent of AIDS. During the news conference, no credence was given to the achievement of the Pasteur Institute. In contrary, Gallo continued to disparage the French discovery, which was not only unethical but also delayed the development of a robust and life-saving blood test to detect HIV.

It did not take long for the French scientists to suspect that Gallo's HLTV-IIIB was simply another name for their LAV because both viruses behaved almost identically during their comparison experiments. Most conspicuously, the genetic similarity between HLTV-IIIB and LAV was 98.3 percent. Statistically speaking, the chances of their being two different viruses from two different patients were almost zero. Even for the same virus, the genetic fingerprints varied from patient to patient to a certain extent, certainly significantly more than 1.7 percent. Therefore, they were exactly the same virus, from the same patient. Gallo and Popovic simply took Pasteur team's LAV, renamed it HTLV-IIIB, and then called it their own discovery. They further took advantage of the virus to develop a blood test for HIV antibodies which would bring financial gains to both the American government and to Gallo himself. In essence, Gallo cheated the Pasteur team of both scientific glory and patent royalties, which would eventually amount to millions.

The Pasteur Institute sued the Americans for misappropriating their virus and stealing their seminal discovery. The disputes for the priority of the discovery of HIV and the patent for the blood test between the Pasteur Institute and NCI went on for three years. In the end, the venerable Jonas Salk volunteered to serve as the intermediary between the two parties. French Prime Minister Jacques Chirac and American President Ronald Reagan were involved to forge a mutually agreeable conclusion in early 1987. Later, after Gallo's misconducts were exposed by the NIH and the US Congressional investigations, a second settlement between the Pasteur Institute and the NIH was reached in 1994, which gave the French more shares of the patent royalties.[4]

The genesis of the two investigations on Gallo's fraudulent scientific conduct traced back to a 1989 article in the *Chicago Tribute*. A Pulitzer Prize-winning journalist, John Crewdson, published a fifty-thousand-word exposé on Gallo's conducts around the discovery of HIV.[5] Two years later, Gallo wrote a book titled *Virus Hunting, AIDS, Cancer, and the Human Retrovirus: A Story of Scientific* Discovery,[6] ostentatiously to rebut Crewdson's accusations by rewriting history. In 2001, Crewdson followed up with a 670-page book, *Science Fictions: A Scientific Mystery, a Massive Cover-up and the Dark Legacy of Robert Gallo*.[7] In the exhaustively researched book, in addition to accusing Gallo of either accidental contamination or outright theft of the Pasteur virus, Crewdson dredged up a lot of dirt on Gallo. For instance, some experts in the virology field still had doubts about the priority of Gallo's discovery of HTLV because he refused to distribute samples until after having received samples from his Japanese competitors. To which Gallo complained: "I would even have to endure a long and trying inquiry within NIH as a result of one journalist's bizarre and obsessively defamatory article,"[8] his own obsession with the media notwithstanding. Apparently, Gallo belonged to a new breed of scientists who did not shy away from promoting himself while promoting science. This would be considered "not gentlemanly" or "bad form" by his predecessors of previous generations. Gallo's love–hate relationship with journalists was reminiscent of that between the media and Princess Diana, a contemporary celebrity known for her HIV/AIDS activism. Princess Diana would court the journalists and fed them information to fuel the fire of controversies surrounding herself and the Royal family. In the end, it was also the media that caused her ultimate demise in Paris in 1997.

Sadly, fallacies did not strike Gallo alone. Montagnier made no significant discoveries before or after he stumbled upon HIV, according to some. Acutely aware that a Nobel Prize was at stake for the discovery of HIV, and that the Nobel Foundation particularly abhors controversies, Montagnier hypocritically put up a façade of reconciliation with Gallo in public, although he seethed privately at every slight by Gallo. In 1990, Montagnier took a leave of his senses and claimed that HIV itself would not cause AIDS without a co-factor such as a bacterial agent. So absurd was his theory that he became a laughingstock in San Francisco when he presented his idea at a conference. Embarrassed, he hastily retreated to France soon after his talk. His proposal echoed a theory perpetuated by Berkeley professor, Peter Duesberg, who erroneously claimed that the HIV did not cause AIDS and that HIV was rather a harmless passenger virus. His theory probably caused millions of unnecessary HIV infections in South Africa, where he advised the government. It made Duesberg a pariah in the scientific community. Once again, in the early 2020, Montagnier spoke out of turn and claimed that the virus that caused COVID-19 was artificially created in a lab by biologists working on an AIDS vaccine. Gallo, who retreated to Maryland, must be laughing at his rival with some satisfaction.

All the drama revealed one thing: scientists are just humans, with same desire for fame and fortune and shortcomings of arrogance and greed. Gallo's thirst for public attention and adulation was half quenched with flying colors. His role appeared, as a villain, in a 1993 movie depicting the history of AIDS, *And The Band Played On*, based on a book of the same title by Randy Stilts published in 1987.[9]

2.1.3 The Nobel Prize

It was highly anticipated that the 1988 Nobel Prize for Physiology or Medicine would go to the discovery of HIV. Undoubtedly it was on the short list of eight potential candidates. But instead, three industrial scientists received the calls from Stockholm. One of them was British pharmacologist James W. Black, for his discovery of beta blockers to treat hypertension. The other two were chemists from Burroughs Wellcome in the United States, George H. Hitchings and Gertrude B. Elion, for their discovery of a range of medicines, including some of antiviral drugs highlighted in this book.[10]

We can only speculate about why the discovery of HIV was not chosen in 1988. At the time, there was no effective treatment for AIDS. While AIDS still a death sentence, it was unsavory to celebrate the discovery of HIV, a deadly virus that caused the disease. In addition, Montagnier, a strong contender for the prize, was not universally loved by his colleagues at the Pasteur Institute. Two living Nobel Laureates from the Pasteur Institute wrote to the Nobel Committee, threatening to return their prizes if Montagnier were to be given the accolade. They declared that never before did a scientist receive a Nobel Prize for such a little thing. Some other colleagues of Montagnier's were no more charitable, pronouncing that Gallo ripped off Montagnier and Montagnier ripped off Barré-Sinoussi.

Yet, the discovery of the HIV was one of the major scientific achievements in the last century, certainly worthy of a Nobel Prize. Another ten years later, in 2008, plenty of AIDS treatments became available and patients stopped dying in droves. The Nobel Committee then saw fit to bestow a prize to the discovery of HIV, which went to Montagnier and Barré-Sinoussi for one-half of the prize. The other half went to German virologist Harald zur Hausen for his "discovery of human papilloma viruses (HPVs) causing cervical cancer" while he was at the University of Freiburg. In the following days, 105 scientists around the world, many of them Italian, wrote a letter to the *Science* magazine titled "Unsung Hero Robert C. Gallo."[11] They argued that Gallo contributed to the discovery of HIV by growing the virus, convincing the world that HIV is the causative agent of AIDS, and developing a blood test. Like Francis Darwin said: "In science, the credit goes to the man who convinces the world, not to whom the idea first occurs." Therefore, Gallo should not go unrecognized. Professor Anders Vahlne at the Karolinska Institute had spent two years conducting a painstakingly thorough investigation before the award. His fair and accurate historical reflection[12] provided a rare window through which to glimpse the inner workings of the Nobel committees as the committee members cannot comment on how they came to their decision. Vahlne stated: "There is *no doubt or controversy* about the fact that the French group was the first to isolate the new virus. This is what the Nobel committees chose to award" because the prize was for the *discovery* of HIV. Vahlne himself actually felt that Gallo deserved to share the prize with Montagnier and Barré-Sinoussi, but he did not have a vote. However, Vahlne based his judgment solely on publications that were part of the public record. He did not factor in the fact that some of Gallo's data were fraudulent in his famous four 1984 *Science* papers, which were subsequently carefully examined by inquiries of the NIH and the US Congress. Maybe this is why we call it a controversy because it is . . . controversial! Everyone has an opinion, and no one is absolutely objective.

What is the fascination, by laymen and scientists alike, with the Nobel Prize? Why is winning the Nobel Prize considered the pinnacle of science and medicine? Gallo, for one, was never bashful to tell anyone who listened that he would do anything to get the Nobel Prize.

When Alfred Nobel set up his will in 1899, the Nobel Prize was to be awarded every year to men or women who made significant contributions in five different disciplines: physics, chemistry, physiology or medicine, literature, and peace. The economics prize was added in 1969. Each award can be given to only three people in any year under any category.

The Nobel Prizes for Physiology or Medicine went pretty well at first, having honored Emil van Behring, Ronald Ross, Ivan Pavlov, Robert Koch, Paul Ehrlich, and many other luminaries in the field. Nevertheless, the 1923 Prize to Frederick Banting and John MacLeod for their "discovery of insulin" was highly controversial.

Nothing was controversial about awarding the discovery of insulin—insulin saved the lives of millions of diabetics immediately after its discovery—a resounding scientific triumph. The debate was over *who* was awarded. The reality was that insulin was discovered by Banting and Charles Best at the University of Toronto in 1921.

In 1920, Banting was a country doctor in London, Ontario, Canada, a small city about 110 miles west of Toronto. At the end of October, an article on diabetes inspired him to look for insulin in the pancreas of dogs. In the summer of 1921, Banting moved to Toronto and obtained support from Macleod, a renowned professor of physiology at the University of Toronto and world expert in carbohydrate metabolism. Before departing to Scotland for his summer vacation, Macleod provided Banting with some laboratory space, several dogs for experiments, and a student assistant, Charles Best. To make the long story short, in the smothering heat of the summer, Banting and Best succeeded in isolating insulin from the pancreas of dogs. Macleod came back from Scotland in September when much of the discovery was already accomplished. Macleod was certainly helpful in improving the process of isolating more and purer insulin. But, in all fairness, the credit should have gone to Banting and Best. Regrettably, the Nobel Foundation's liaison to investigate the circumstances of the discovery was August Krogh, the 1920 laureate from Denmark, who was convinced by Macleod to believe his version, which grossly exaggerated MacLeod's contributions. When the Karolinska Institute announced that MacLeod would share the 1923 Nobel Prize with Banting, an outraged Banting immediately announced that he was going to share one-half of his prize money with Best. The messy aftermath was a stain on Canadian science and a big embarrassment to the Nobel Foundation.[13] By coincidence or not, no Canadian would receive a Nobel Prize in Physiology and Medicine for nearly a century after that. It was not until 2020, when Michael Houghton at the University of Alberta shared the prize with two Americans—Harvey J. Alter and Charles M. Rice—for their "discovery of hepatitis C virus" that Canada was again honored.

Three years after the Banting/MacLeod controversy, the Nobel Committee committed another error in their selection of Johannes

Fibiger as the 1926 awardee for "his discovery of the *Spiroptera carcinoma*." Danish pathologist Fibiger studied bacteriology in his youth under the famous Robert Koch. He later took up a position as the Director of Copenhagen's Institute of the Pathological Anatomy. In 1907, he performed pathological studies on mice in a Copenhagen sugar refinery. The mouse's stomach walls were shown to have unusual growth that seemed to be malignant, a sign of papillomatous stomach tumors. Upon closer scrutiny, the mouse stomachs were infected with a tiny nematode worm, *Spiroptera*. As it so happened, the sugar refinery was infested with cockroaches that were brought as stowaways on the sugar ships from the Danish Virgin Islands. The cockroaches carried those nematode worms as parasites. The mice developed "stomach cancer" by eating those cockroaches with the *Spiroptera* worm parasites. In 1913, Fibiger proposed that cockroaches became infested by eating mice excrement containing parasitic eggs and that mice were then re-infected by eating larvae-lade cockroaches. Irritation or chemical reactions from the nematodes caused stomach cancers, which bore resemblance to frequently occurring human stomach cancers.

The Karolinska Academy of Sciences was mightily impressed by the august pathologist who finally solved the mystery of the etiology of human stomach cancers. They promptly bestowed the 1926 Nobel Prize to Fibiger. Sadly, Fibiger died of an aggressive colon carcinoma merely six weeks after the award ceremony in Stockholm. More disconcertingly, other scientists could not reproduce his results. It turned out that the mice in question suffered from a vitamin A deficiency because they ate mostly cockroaches in the sugar refinery and white breads in Fibiger's laboratories. In addition, it was further revealed that the stomach cancer of the mice that Fibiger observed was closer to pre-malignant thickening of the stomach wall, not really a cancer malignancy.[14] Stunned by the *faux pas*, the Nobel Foundation awarded no cancer-related research in the next forty years.

Macleod in 1923 and then Fibiger in 1926, two debacles in a row had stunned the Nobel Committees. Much closer scrutiny has since been applied to both the researchers and the awardees. In the movie *Beautiful Mind*, an emissary was sent to the Princeton University to check on John Nash's mental state before announcing his Nobel Prize for Economics in 1994.

During the heydays of the discovery of HIV, the Pasteur team was profoundly disturbed by Gallo's incessant denigration of their important contributions. The genuine or perceived slight of French scientists by "Imperial American" colleagues was not an isolated incident. While Montagnier's scars were still fresh, another controversy between the French and Americans flared up again in 1989. Michael Bishop and Harold Varmus received the Nobel Prize that year for their "discovery of cellular origin of retroviral oncogenes." Their French postdoctoral fellow, Dominique Stehelin, bitterly and publicly accused them not giving him enough credit. He was also under pressure from his French superior, who was determined to get credit for France. Stehelin appeared on French television offering his laboratory notebooks for verification of his claims. He carried out the experiments from start to finish and he was the first author of that important paper on *Nature*.[15] In reality, because he was instructed to do what he did, it was difficult to justify his sharing the Prize. Even Jan Lindsten, the secretary of the Nobel Committee, told *Science* that the committee thought that Bishop and Varmus were the key persons in the discovery.[16]

Quite ironically, it was widely believed that Robert Weinberg should have shared the 1989 Nobel Prize with Bishop and Varmus. Weinberg made significant contributions to the discovery of retroviral oncogene *RAS* and the tumor suppressor gene *Rb*, respectively.[17] Incidentally, today KRAS-G12C has become one of the hottest drug targets in oncology, even though KRAS had been considered undruggable for decades. It is now druggable using the targeted covalent inhibitor strategy.

2.2 The First Ray of Hope

In the darkest days of the HIV/AIDS epidemic, the emergence of azidothymidine (AZT, azidothymidine) was the first glimmer of hope for the patients. AZT was the first agent licensed for the treatment of AIDS, just four years after the identification of the HIV as the etiologic agent. AZT belongs to a class of compounds known as nucleosides that are the building blocks of DNA and RNA. Each nucleoside contains a deoxyribose core attached to one of the four natural bases, which may be adenine (A), cytosine (C), guanine (G), or thymine (T) (see Figure 2.3). In some cases, uracil (U) substitutes for thymine (T).

2.2.1 Nucleoside antiviral drugs

Nucleoside antiviral drugs saved the day when AIDS was a virtual death sentence in the early 1980s when there was no effective drug at all. AZT is a substrate analog of thymidine and the incorporation of AZT triphosphate into the growing DNA chain, which results in termination. But to appreciate the emergence of AZT, we must trace back to the first nucleoside antiviral drug, idoxuridine, discovered by William H. Prusoff.

2.2.1.1 The godfather of modern antiviral chemotherapy

In the 1950s, George H. Hitchings started to systemically investigate purine and pyrimidine analogs as potential drugs. Inspired by Hitchings, William H. Prusoff at Yale University synthesized idoxuridine (IdU) in 1959, which became the first small-molecule antiviral drug.[18] Although idoxuridine was too toxic to be given systemically, it was applied topically to treat eye and skin infection caused by the herpes simplex virus (HSV). Herbert Kaufman, a resident at Massachusetts Eye and Ear Infirmary in Boston, studied idoxuridine in the early 1960s. Using rabbits as an animal model,

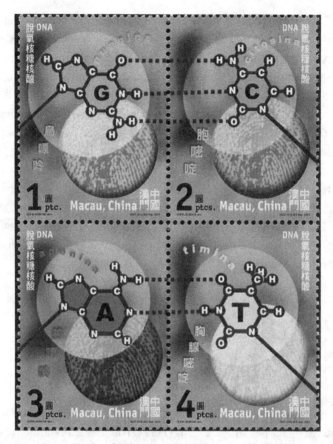

Fig. 2.3 Macao Post and Telecommunications Bureau has granted permission for using the image of postage stamp "Science and Technology—Composition and Construction of DNA"

Kaufman discovered that idoxuridine was an effective treatment of herpes simplex keratitis, the most common infection of the human cornea. Caused by herpes virus infection, herpes simplex keratitis is associated with significant morbidity of eyes. Without treatment, a herpes virus infection of the human eye may eventually lead to impaired vision and even to blindness.

While its mechanism of action is not completely elucidated, idoxuridine is most likely phosphorylated first by kinases in either virus or normal cells. It is then sequentially transformed to the corresponding nucleotide monophosphate, nucleotide diphosphate, and nucleotide triphosphate. When a phosphate is attached to a nucleoside, it becomes a nucleotide. Nucleoside triphosphate is the real active drug with two fates. On the one hand, when interacting with viral DNA polymerase, it terminates DNA replication and exerts antiviral activities (good). On the other hand, when interacting with host cellular DNA polymerases, cytotoxicity (toxicity against host cells) and mitochondrial toxicity ensued (bad).

The emergence of idoxuridine opened a floodgate of ribonucleoside antiviral drugs. It was followed by four additional "me-too" drugs:

- idoxuridine (IdU, Herplex) in 1962,
- trifluorothymidine (TFT, Viroptic, GSK) in 1980,
- ethyldeoxyuridine (EdU) in 1985,
- bromovinyldeoxyuridine (brivudine, Zostex) in 2001, and
- telbivudine (Tyzeka, Novartis), 2006.

Bromovinyldeoxyuridine (Brivudine) is used to treat shingles caused by herpes simplex virus (HSV), whereas telbivudine is also a synthetic thymidine nucleoside analog put on the market by Novartis in 2006 to treat hepatitis B infection.

Not only was Prusoff a pioneer of the first antiviral drugs, he also contributed, in the late 1980s, to the discovery of d4T (stavudine, Zerit), one of the early nucleoside reverse transcriptase inhibitors to treat AIDS immediately after AZT was found to be effective. Prusoff and his Yale colleague Tai-Shun Lin, with the help from Raymond Schinazi at Emory University, characterized d4T's HIV inhibitory activity. Independently, Erik De Clercq at Rega Institute of for Medical Research at the University of Leuven in Belgium and a team in Japan led by Yoshiaki Hamamoto at Yamaguchi University also made similar observations on d4T's antiviral activities. The three groups at Yale, Leuven, and Tokyo almost simultaneously making the same discovery engendered some international animosity in the old days. Even though d4T was not as potent as AZT, it had a much better safety profile and was more bioavailable. Because d4T's structure had long been known, a *composition of matter* patent was impossible. But Yale secured a *method of use* patent. Successful clinical development of d4T carried out by the neighboring Bristol-Myers Squibb in Wallingford, Connecticut garnered the FDA approval in 1994. It went on to become one of the most popular anti-HIV drugs to be used worldwide at the time. The drug generated a huge revenue in royalties for Yale, far surpassing the amount generated by all its other licensed medicines combined. In fact, Yale's patent royalty grew ten-fold from the d4T (Zerit) royalty alone. Prusoff, the gentleman scientist, was not about money, but about saving lives. When Yale students and Doctors without Borders called on the university to make Zerit available at low cost for poor countries, Prusoff actively joined in the campaign. He said: "We were not doing this to make money, we are interested in developing a compound that would be a benefit to society."[19] Today, such gentleman scientists are few and far between. Greed has permeated everywhere.

2.2.1.2 AZT

Because HIV is a retrovirus, blocking its reverse transcriptase is unique, thus providing certain selectivity against virus cells and infected cells over healthy host cells. In October 1984, the newly discovered retrovirus HIV was the subject of a series of seminars given at Burroughs Wellcome by Françoise Barré-Sinoussi, Robert Gallo, and Sam Broder, which ignited their interests in the field. In February 1985, Sam Broder at the NCI and two of his post-doctoral fellows, Hiroaki "Mitch" Mitsuya and Robert Yarchoan, commenced a screening program of promising drug candidates as a potential treatment of AIDS. They solicited submissions from pharmaceutical companies to supply potential antiviral drugs. About fifty companies submitted approximately 180 compounds. Many nucleosides and nucleotides were submitted. Most of them were initially prepared as potential anti-cancer drugs but were known to affect either DNA or RNA synthesis. From the screening, only one compound showed an ability to impair or deplete the functions of HIV's reverse transcriptase activities using an immortalized human T_4 cell culture assay newly developed by Mitsuya. That compound was AZT, one of the twelve compounds submitted by Janet Rideout, a nucleoside chemist at Burroughs Wellcome (now known as Glaxo SmithKline, GSK). AZT was initially synthesized in 1964 by Jerome Horowitz at the Detroit Institute of Cancer Research of the Michigan Cancer Foundation as a potential treatment for cancer but it was not pursued further because of a lack of activity against animal cancers. Ten years later, Wolfram Ostertag at the Max Planck Institute tested AZT and found it active against a leukemia virus known as *Friend virus*, a retrovirus similar to HIV. That was actually the justification for its selection to be among the first compounds screened for HIV inhibition. But back in 1974, nobody paid much attention because Friend virus was merely an exotic virus, despite it auspicious name.

Once AZT's inhibitory activity of HIV reverse transcriptase was observed, its selectivity was investigated. AZT-triphosphate blocked the synthesis of DNA by inhibiting reverse transcriptase about one hundred times better than it inhibited the synthesis of DNA by the host cell DNA polymerase in the cell nucleus. AZT inhibited HIV replication at concentrations about one thousand-fold less than it took to inhibit the replication of the host cell lymphocytes. In other words, AZT killed the viral target T-cells faster than normal healthy cells, thus providing a therapeutic window. Armed with those data, Burroughs Wellcome and NCI swiftly completed an an Investigational New Drug Application

(IND) for the FDA, so they could give the drug to patients. The first-to-patient (FIP) took place in July 1985 when AZT was given to an AIDS patient, Joseph Rafuse, who was a furniture salesman from Massachusetts. After six weeks, his T-cell count had significantly increased. The Phase I clinical trials enrolled thirty-five patients. The purpose for the Phase I clinical trials was largely to gauge the drug's toxicity and pharmacokinetics: how human body responded to the drug. After several months, AZT raised the CD4 counts in AIDS patients and had a reasonable safety profile.

With promising Phase I data, a double-blinded, placebo-controlled, randomized (known as the gold standard of clinical trials) Phase II trial began with 282 AIDS patients. About one-half of them were given AZT and the other half were given a placebo. The goal was to gauge the efficacy of the drug; that is, did the drug work to treat the disease? At the end of the trial, the answer was a resounding yes! There was one death out of 145 patients on AZT while there were nineteen deaths among those who received the placebo. The trial was terminated in September 1986 by the Drug Safety Monitoring Board so that patients on the placebo could be switched to AZT. Otherwise, it would be unethical. In terms of AZT's mechanism of action, it is an HIV nucleoside reverse transcriptase inhibitor (NRTI), serving as a chain-terminator of viral DNA synthesis. The only structural difference between thymidine and AZT is that a hydroxyl group is replaced by an azide group. After phosphorylation by kinases, the resulting AZT-triphosphate "pretends" to be thymidine-triphosphate and is incorporated by the reverse transcriptase into viral DNA. At the end of a growing DNA chain, it prevents additional DNA chain growth. This is why nucleoside antiviral drugs, such as AZT and d4T, are known as DNA chain terminators.

AZT was approved by the FDA in March 1987 for the treatment of AIDS patients. It only took twenty-five months from observation of laboratory activity against HIV to its approval.[20]

2.2.1.3 *Nucleoside reverse transcriptase inhibitors*

Regrettably, AZT is toxic to bone marrow and can cause anemia, headaches, and nausea. In addition, HIV often mutates when challenged by a single therapeutic agent, creating drug-resistant strains of the virus that foil treatment. Consequently, only patients with advanced-stage HIV disease were given the drug. Safer therapeutic agents were desperately needed to provide a combination therapy to fully suppress HIV and to prevent the development of drug resistance.

After AZT was approved by the FDA in 1987, at least seven "me-too" and "me-better" nucleoside reverse transcriptase inhibitors emerged on the market for treating AIDS:

- zidovudine (AZT, azidothymidine, Retrovir, Burroughs Wellcome) in 1987;
- didanosine (ddI, Videx, BMS) in 1991;
- zalcitabine (ddC, Hivid, Hoffmann LaRoche) in 1992, but withdrawn in 2006;
- stavudine (d4T, Zerit, BMS) in 1994;
- lamivudine (3TC, Epivir, GSK) in 1995;
- abacavir (Ziagen, GSK) in 1999;
- tenofovir disoproxil (Viread, Gilead) in 2001; and
- emtricitabine (FTC, Emtriva, Gilead) in 2003

Following AZT's footsteps, Sam Broder at the NIH quickly identified ddI and ddC as similar HIV reverse transcriptase inhibitors. Both ddI and ddC were already commercially available at the time; therefore, there were no patent issues with which to contend, and they were approved by the FDA in 1991 and 1992, respectively. For ddC, its use was limited due to its toxicities, which ultimately led to its withdrawal from the market at the end of 2006. In fact, AZT, ddC (zalcitabine), and d4T (stavudine) were all synthesized as potential cancer drugs by Jerome Horowitz in the 1960s. Regrettably, there were many problems associated with poor tolerability and the emergence of resistance with the use of these drugs, especially ddI and ddC. Better drugs were still needed.

Then came lamivudine (3TC). In 1988, Bernard Belleau at BioChem Pharma, a Quebec-based drug company, first synthesized BCH-189. It was a mixture of two enantiomers of an oxathiolane nucleoside and one of them was lamivudine (3TC).[21] One feature unique to lamivudine was that it had a sulfur atom on its deoxyribose scaffold. It belonged to a novel class of oxathiolane nucleosides. While studying its toxicity, Yung-Chi (Tommy) Cheng at Yale found that lamivudine had reduced side effects when used in combination with AZT. BioChem obtained a patent on the drug in 1991. Licensing the right from BioChem Pharma, Glaxo Wellcome developed lamivudine through clinical trials, gained marketing approval in 1995, and sold it with the trade name Epivir. Because Glaxo Wellcome already owned AZT, it made sense for them to combine AZT with lamivudine as a single-tablet regimen. *Combination drug therapy* would become the hallmark of modern AIDS drugs.

Virologist Raymond Schinazi, Professor of Pediatrics at Emory University and a former student of Prusoff, attended the Fifth International Conference on AIDS in 1989. He was intrigued by BioChem's poster on BCH-189, which was reported to exhibit good anti-HIV activity with no apparent cytotoxicity. Schinazi relayed the news to his colleague at Emory, Dennis Liotta, a professor in the chemistry department. Working with his postdoctoral fellow, Woo-Baeg Choi, Liotta made two significant contributions to the oxathiolane nucleoside drugs. First, they discovered that one of the two enantiomers, the left-handed isomer, was less cytotoxic than the right-handed isomer even though they were almost equally potent. They subsequently developed methods to make the desired enantiomer by either stereoselective synthesis or chiral resolution. Second, among many similar analogs that they prepared, they discovered that the drug with an additional fluorine atom was superior to BCH-189. The fruit of their innovation was emtricitabine (FTC), which has been sold since 2003 by Gilead Sciences with the trade name Emtriva. Now, every card-carrying, self-respect medicinal chemist today would have done it regarding both enantiomers and fluorination at the first sight of the molecular structure of BCH-189. But remember, this was thirty years ago and even separating two enantiomers of *existing drugs* was a novel concept. For instance, a biotech company, Separacor Inc., was ostentatiously founded in 1984 with a mission of separating enantiomers of drugs already on the market and developing the better enantiomer with either higher efficacy or better safety profile, or both. For instance, Separacor separated the two enantiomers of insomnia drug zopiclone (Imovane) and marketed its left-handed isomer as eszopiclone (Lunesta).

Due to priority issues and the drug company's M&A (Merger and Acquisition), Emory engaged in a long, convoluted, acrimonious patent dispute involving BioChem Pharma, Burroughs Wellcome, later GlaxoWellcome and Glaxo SmithKline, Triangle Pharmaceuticals, Shire Pharmaceuticals, and Gilead Sciences. But it was worthwhile because Emory obtained over $500 million in royalties and the three inventors—Liotta, Schinazi, and Choi— were entitled to forty percent of the royalty. Money aside, the discoveries saved many AIDS patients' lives. Over ninety percent of HIV-infected persons on therapy in the United States take a drug containing either lamivudine or emtricitabine.[22]

Excitingly, as of 2020, Merck was developing a nucleoside reverse transcriptase inhibitor islatravir (MK-8591). Merck acquired MK-8591 in 2012 from Yamasa, a Japanese company best known for making soy sauce! Because islatravir has an extremely long intracellular half-life (>120 h), it will be one of the long-lasting antiretroviral drugs, which will increase patient compliance exponentially.

The initial nucleoside reverse transcriptase inhibitors were used as monotherapies, and they were not very efficacious in slowing down the virus progression. Further, drug resistance developed within several months. They also had poor oral bioavailability and had to be given in large doses. This is why AIDS patients had to take many gigantic pills when they were first on the market. Today, AZT and d4T are largely superseded by newer nucleoside reverse transcriptase inhibitors, including lamivudine, emtricitabine, carbocyclic nucleosides abacavir, as well as acyclic nucleotide analogue tenofovir (which is marketed as a prodrug, tenofovir disoproxil fumarate, with the trade name Viread). More excitingly, with the discovery of the HAART (cocktail drugs) in the mid-1990s, the AIDS pandemic was finally under control. The cocktail drugs contain a nucleoside reverse transcriptase inhibitor, an HIV protease inhibitor, and a non-nucleoside reverse transcriptase inhibitor, which happens to be the topic of the following discussion.

2.2.2 Non-nucleoside reverse transcriptase inhibitors

When we look for drugs, we strive to increase the drug's efficacy and minimize its toxicities. As the Hippocratic Oath states:

Primum non nocere:
First, do no harm!

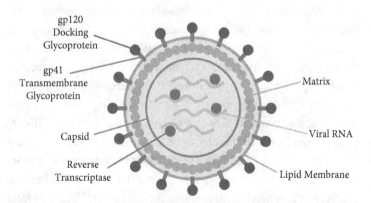

Fig. 2.4 The structure of the HIV, diagram by Alexandra H. Li

Non-nucleoside reverse transcriptase inhibitors have less cytotoxicity in comparison to their nucleoside counterparts. Let's discuss why this is.

Nucleoside reverse transcriptase inhibitors inhibit the HIV reverse transcriptase as a chain terminator by mimicking a natural substrate of the enzyme, thymidine. However, they just as well mimic those nucleosides naturally present in the host cells. This might be why they have certain side effects: because it lacks selectivity between the bad enzyme (HIV reverse transcriptase) and the good endogenous human enzymes in host cells. Therefore, nucleoside reverse transcriptase inhibitors harm both virus cells and host cells with similar ferocity. It is therefore not that surprising that toxicities are a major drawback of these drugs.

The trick is: can we find a drug that will be active against the virus but will have little or no toxicity to the mitochondrial DNA?

Imagine, if we could find drugs that only blocked the enzymatic activity unique to retroviruses that were not found in normal human cells. We would get more selective drugs with decreased cytotoxicity. *Non-nucleoside* reverse transcriptase inhibitors (NNRTIs) are such drugs. Boehringer Ingelheim was the first to reach the finish line with nevirapine (Viramune), which was approved by the FDA in 1996.

In 1990, Boehringer Ingelheim used a low through-put screening that tested merely 100 compounds per week. Fortuitously and fortunately, one hit emerged after testing only about six hundred random compounds. That hit compound was pirezenpine (Gastrozepin), which was previously synthesized as an M_1-selective anti-muscarinic compound for the anti-ulcer drug discovery program. Because it was weakly active to inhibit the activity of HIV reverse transcriptase, the task fell upon the medicinal chemistry team to make the hit better: more potent, more bioavailable, and more efficacious with fewer toxicities. The chemistry team was composed of several future prominent chemists including Karl Hargrave, John Proudfoot, and Julian Adam. To make a long story short, the team eventually came up with two potential clinical candidates in due course. One had an N-ethyl group and the other (nevirapine) had an N-cyclopropyl substituent. The team wisely chose nevirapine to develop because it was more resistant to N-dealkylation than the corresponding N-ethyl compound, even though it was slightly more potent.[23] Indeed, because the cyclopropyl motif has partial aromatic properties, it is thus more resistant to liver metabolism. Today, it is routine to choose the cyclopropyl group over other simple alkyl groups to minimize metabolism. In fact, abacavir, another nucleoside reverse transcriptase inhibitor, also takes advantage of the magic cyclopropyl functionality.

The second non-nucleoside reverse transcriptase inhibitor on the market was efavirenz (Sustiva), discovered by DuPont-Merck Pharmaceuticals, a joint venture between DuPont Central Research and Merck. Its discovery followed a similar route to Boehringer Ingelheim's nevirapine. Merck obtained a hit from screening over one hundred thousand compounds—not as lucky as Boehringer Ingelheim's screening campaign. In addition to being weak, the hit compound had a thiourea group, a well-known toxicophore as thioureas tended to cause liver toxicities. Switching the thiourea to urea minimized the toxicity and the addition of a cyclopropylacetylynyl group significantly boosted the potency. Although its tertiary methyl group was readily metabolized, replacing the methyl with a trifluoromethyl successfully blocked the metabolism. The fruit of their labor was efavirenz (Sustiva), which was approved by the FDA in 1998.[24]

Nevirapine and efavirenz, along with Upjohn's delavirdine, are considered the *first-generation non-nucleoside reverse transcriptase inhibitors*. They are generally well absorbed and have good safety profiles. But the rapid emergence of drug resistance, such as L100I and K103N mutations, dramatically reduced their potency. The HIV reverse transcriptase has this cunning ability to mutate one or more of its amino acids to a different one. For instance, L100I is when leucine (L), at the 100th position of the reverse transcriptase enzyme, is converted to isoleucine (I). In the same vein, K103N mutation is when lysine (K), at the 103rd position of the enzyme, is converted to asparagine (N). Having mutated, the enzyme does not allow the original inhibitor to bind as tightly any longer, and thus the drug loses potency. This often results in poor clinical patient compliance.

Therefore, the first-generation non-nucleoside reverse transcriptase inhibitors are mostly eclipsed by the second-generation counterparts which possess a high genetic barrier to resistant clinically relevant mutations. They accomplish such a feat by their *structurally intrinsic flexibility* and readily adapt conformations to adjust to the enzyme's binding pocket. You see, not only does *flexibility* make you a better person—it also makes better drugs.

The second-generation non-nucleoside reverse transcriptase inhibitors include etravirine (Intelence), rilpivirine (Edurant), and doravirine (Pifeltro).

On November 5, 1986, De Clercq had an eight-hour discussion with his compatriot, Paul Janssen, the legendary drug hunter whose name was associated with more than fifty marketed drugs. In 1987, they began a collaboration to find the ideal AIDS treatments. They carried out a rational screening program of six hundred well-chosen compounds from the Janssen library using an MT4 cell-based anti-HIV assay. Rudi Pauwels, who used to work at Janssen but now worked at Leuven after his retirement, developed the assay using MT4 cells infected with the LAV virus provided by Montagnier. That batch of MT4 cells was also infected with the HTLV-III strain of the virus procured from Gallo. Little did they know that those two viruses were actually exactly the same. The Leuven and Janssen teams discovered a hit compound trivirapine (TIBO), from which Janssen arrived at a class of diarylpyrimidines. This would be the starting point of almost all following non-nucleoside reverse transcriptase inhibitors. In collaboration with Tibotec in Belgium and Microbicide in the United States, two decades of intensive multidisciplinary research involving medicinal chemistry, virology, crystallography, molecular modeling, toxicology, and other disciplines eventually led to the discovery of etravirine (Intelence) and rilpivirine (Edurant), respectively. Unlike the first-generation non-nucleoside reverse transcriptase inhibitors that have a "butterfly-like" shape, etravirine and rilpivirine have a "horseshoe-like" conformation that allows "wiggling" and "giggling." They fit snugly within the non-nucleoside binding pocket site of the HIV reverse transcriptase located at some distance from the catalytic site, so they are "allosteric" inhibitors. They are very active against mutant strains because they are flexible enough to allow tight binding even though the enzyme has mutated.[25]

All in all, seven non-nucleoside reverse transcriptase inhibitors are currently on the market:

- nevirapine (Viramune, Boehringer Ingelheim) in 1996,
- delavirdine (Rescriptor, Upjohn) in 1997,
- efavirenz (Sustiva, Merck/BMS) in 1998,
- etravirine (Intelence, Tibotec/Janssen) in 2008,
- rilpivirine (Edurant, Tibotec/Janssen) in 2011,
- doravirine (Pifeltro, Schering Plough/Merck) in 2018, and
- elsufavirine (Elpida, Viriom) in 2018 in Russia.

2.3 HIV Protease Inhibitors

Before we embark on the journey of the discovery of HIV protease inhibitors, it is an opportune time for us to look at the virus' lifecycle. This will give us a perspective of HIV protease's functions in the virus' replication.

2.3.1 HIV protease

HIV protease is a virus-specific enzyme involved in processing and maturation in HIV's lifecycle. Protease inhibitors, in turn, function by blocking further viral processing and maturation.

An HIV by itself is a virion, which enters the blood stream through sexual intercourse, blood transfusion, or the vertical route (from mother to infant). For the virion to enter the host cell, its envelope spike protein gp120 makes contacts with CD4 on the T-cell membrane. Here, gp stands for *glycoprotein* on HIV's envelope protein. The process of making contact is called tropism. Once contact is made between gp120 and CD4, the virion "injects" its RNA genetic information into the host cell with the help of another envelope glycoprotein, gp41. But the RNA does not travel alone. Its travel companions include three enzymes serving as its accomplices: reverse transcriptase, protease, and integrase. Inhibition of reverse transcriptase and protease has proven to be the most successful approach to effective therapies and most of the approved AIDS drugs target these two enzymes.

Once HIV has gained entry into the cell, the virus sheds its protein coat, releasing its genetic information of the RNA strand along with the reverse transcriptase into the cytosol. Inside the cytosol, the virus can take advantage of the availability of substrates to reversely transcribe the RNA genome into a double-stranded DNA copy using reverse transcriptase.

HIV protease, the enzyme that HIV needs to make new virus cells, is responsible for processing a couple of polyprotein gene products into mature and functional proteins. In essence, some large viral proteins must be broken down to smaller proteins with regulatory functions by the protease. It plays a critical role in virus replication and is the best characterized of the virus' proteins, both functionally and structurally. It is an aspartyl protease similar to renin in humans in its catalytic mechanism. Other human aspartic proteases include pepsin, gastricsin, and cathepsins. The HIV protease is formed by the symmetrical dimerization of monomeric subunits, each of which contributes a catalytic aspartate residue. Therefore, it is not surprising that many drug firms initially screened their old renin inhibitors (for regulation of blood pressure) to look for a hit for HIV protease inhibitors. Protease inhibitor design was markedly facilitated by the availability of X-ray crystallographic structures that allowed direct observation of inhibitors bound to the enzyme.

Although protease acts at the last step of the virus life cycle, protease inhibitors were discovered shortly after nucleoside reverse transcriptase inhibitors. The era of HIV protease inhibitor drugs began in 1996 with the introduction of saquinavir, indinavir, and ritonavir as the first-generation HIV protease inhibitors.

2.3.2 First-generation HIV protease inhibitors

Roche's saquinavir (Invirase) was the first HIV protease inhibitor on the United States market. Back in 1986, Roche undertook an ambitious international collaboration to tackle the HIV protease. The chemistry team in Welwyn, England was led by Ian Duncan and Sally Redshaw. After choosing a colorimetric assay as their *in vitro* assay, chemists designed inhibitors using the "transition-state mimic" strategy, which was previously highly successful in producing potent renin inhibitors. Because the transition state of a reaction has the highest energy for the reaction thermodynamics, a transition-state mimic is capable of inhibiting the functions of the enzyme. Saquinavir relied upon the hydroxyethylene moiety as an isostere of the tetrahedral transition state, a design element originating from the extensive exploration of inhibitors of renin, another mammalian aspartyl protease. The chemists soon achieved an important milestone by defining the smallest peptide mimetic with which they could achieve an acceptable level of inhibition. They found that a tripeptide was ideal in terms of both potency and bioavailability. Consequently, tripeptides became a common theme in many protease inhibitors to follow. The Roche team made a heroic effort in fine-tuning the tripeptide, exploring their lead compound systematically by modifying each amino acid residue one by one. Their hard work paid off in 1991 when the team arrived at saquinavir, which became the first HIV protease inhibitor on the market for the treatment of AIDS when it was approved by the FDA in December 1995.

Although saquinavir, a peptomimetic, was metabolized easily, it was later found more beneficial to co-administrate with another protease inhibitor, ritonavir by Abbott, as a pharmacoenhancer. The combination could boost the plasma level of saquinavir significantly because ritonavir inhibited the cytochrome P450 3A4 (CYP3A4) enzyme in the liver that degraded saquinavir. CYP3A4 is the most abundant CYP isozyme in the liver. Moreover, the availability of saquinavir marked the beginning of combination therapy as cocktail drugs by introducing protease inhibitors to the clinic.[26]

Abbott's ritonavir (Norvir) was the second protease inhibitor on the market. The circumstances under which ritonavir was discovered were quite unique. The Abbott team was led by X-ray crystallographer John Erickson and medicinal chemist Dale Kempf. While it was rumored that Merck had thirty chemists on their HIV protease inhibitor project, Kempf had three. Instead of screening their renin inhibitors like most drug firms did, Abbott took advantage of Erickson's X-ray crystallography work on the HIV protease, which would prove to be instrumental to their drug design. Integrating structure-based drug design (SBDD) and traditional medicinal chemistry, they prepared a series of symmetry-based inhibitors to match the C_2-symmetric nature of the HIV protease. Kempf dubbed their inhibitors *molecular peanut butter* because they bound to both sides of the enzyme. Using that approach, they arrived at pyridine-containing A-77003, a tetrapeptide. Although A-77003 was potent in binding and cellular assays, it was not bioavailable due to extremely high human biliary clearance. By reducing the molecular weight and replacing the existing amino acids with more soluble ones, they achieved an increase in bioavailability. They finally hit the jackpot when they identified that the pyridine termini were oxidized into *N*-oxide by hepatic cytochrome P450. Simply replacing the pyridines with metabolically more robust thiazoles and a little fine-tuning gave rise to ritonavir, with a bi-oavailability of seventy-eight percent in comparison to twenty-six percent for the pyridyl analog A-77003. In March 1996, Abbott's ritonavir (Norvir) won approval from the FDA.[27]

Abbott's discovery of ritonavir was certainly impressive. But two ensuing events after its discovery were equally, if not more, intriguing. One involved its crystalline forms and the other led to cobicistat, a ritonavir analog discovered by Gilead as a *bioavaila-bility booster*.

After gaining the FDA's approval to sell ritonavir with the trade name Norvir in 1996, Abbott manufactured the drug's active pharmaceutical ingredient (API) with the original crystalline form I. Because ritonavir was not bioavailable in its solid state, it was formulated as either an oral solution or semi-solid capsule, both in an ethanol–water-based solution. In early 1998, after two years on the market, things suddenly went wrong in the manufacturing of API: many lots started to become a new, different, and extremely in-soluble crystalline form II, even though form I was not that soluble itself to start with. There was an extensive intermolecular hydrogen bond network for the new crystalline form II, which was thermo-dynamically more stable than the original crystalline form I. Once there was a crystal seed of form II, no matter how minute and mi-croscopic the seed was, Mother Nature would invariably convert the API to the thermodynamically more stable form II. A team of scientists who had been exposed to the crystalline form II visited Abbott's manufacturing facility in Italy to investigate ritonivir's bulk manufacturing process. Afterward, a significant amount of crystalline form II started showing up in bulk drugs during the manufacturing process done at the plant. The situation really put Abbott at a marketing crisis. It took a Herculean effort of Abbott's process chemists to figure out a way to make only the original form I. To prevent the formation of form II, the semi-solid capsule, or oral formulation, required refrigeration prior to use, among other measures to be taken.[28]

Another story following ritonavir's success took place at Gilead Sciences in Foster City, California. Intriguingly, while many other protease inhibitors are metabolized by CYP3A4, ritonavir is a potent mechanism-based inhibitor of both CYP3A4 and P-glycoprotein (Pgp), which shuttles the drug outside the cell. Therefore, it could be used to increase plasma levels of co-dosed protease inhibitors or any other drugs that are metabolized by CYP3A4. As a result, dual protease inhibitor therapy has proven to be a powerful regimen in terms of efficacy and minimizing drug resistance. Low doses of ritonavir are now used primarily as a pharmacokinetic (PK) booster of HIV inhibitors, also known as a PK enhancer or pharmacoenhancer. A pharmacoenhancer, which itself is often not active against the therapeutic target, but it can inhibit the enzyme that metabolizes the active drug. Therefore, a pharmacoenhancer can improve the PK profiles of an antiviral drug to achieve adequate *trough concentrations* at lower dosage and with less frequent dosing.[29] The trough concentration is the lowest concentration reached by a drug before the next dose is administered.

Gilead worked with Japan Tobacco on the HIV integrase inhibitor elvitegravir (Vitekta, 2014). Regrettably, despite being a potent integrase strand transfer inhibitor (INSTI), elvitegravir was extensively metabolized, primarily by CYP3A4 in both liver and intestines, making a once-daily regimen impossible. They obtained a good boost of drug exposure when given together with ritonavir, although using a sub-therapeutic dose of ritonavir had the potential to cause drug resistance against other protease inhibitors. Gilead decided to make a drug that would only boost bioavailability by inhibiting CYP3A4-mediated metabolism but was completely devoid of HIV protease inhibition. Meanwhile, a high aqueous solubility and good physiochemical properties were desirable to facilitate drug formulation. Ritonavir had a poor aqueous solubility. Under the leadership of Manoj Desai, head of Medicinal Chemistry, a team led by Lianhong Xu arrived at a potent, orally bioavailable, selective CYP3A4 inhibitor, cobicistat, using ritonavir as their starting point. The secret to their success was the removal of a hydroxyl group on ritonavir because it mimicked the transition state of amide hydrolysis through hydrogen bond with the two amino acids at the catalytic domain in the active site of the HIV protease. The molecule without the key hydroxyl group had little binding to the HIV protease. Cobicistat was approved by the FDA in 2012 and Gilead sold it under the trade name Tybost, which has been used in combination with a variety of antiretroviral drugs.[30]

In drug discovery, drug–drug interaction (DDI) occurs when two drugs are metabolized by the same isoform of the liver CYP enzyme. They are often regarded as a liability for good reasons. For example, Bayer's statin cerivastatin (Baycol) caused severe liver damage when it was given with fibrate drugs to lower cholesterol. Since both are metabolized by CYP3A4, the liver does not have enough CYP3A4 enzyme to metabolize both drugs at the same time, thus causing toxicities. But with pharmacoenhancers, such as ritonavir and cobicistat, we can turn a drug's liability into an asset. While the pharmacoenhancers keep the liver CYP enzymes busy, the antiviral drugs can focus on killing viruses without watching their own back. Cobicistat can potentially make a once-daily, protease inhibitor-containing single tablet regimen possible.

Merck's indinavir (Crixivan) was the third protease inhibitor on the market.[31] Merck began their research on HIV protease inhibitors in 1986 with Irving Sigal, a senior director in the Department of Molecular Biology, as the project champion. They initially also screened their renin inhibitors for HIV protease inhibition and then carried out rational drug design by taking advantage of the known crystal structure of HIV protease, first discovered by Merck and refined by the NIH scientists. In 1990, chemist Wayne Thompson arrived at L-689,502, which was active in inhibiting the HIV protease but was devoid of renin activity. Unfortunately, it was not bioavailable and was only effective by injection. By that time, Roche's saquinavir surfaced in the literature as a viable oral drug. Inspired by saquinavir's success, Joseph Vacca successfully incorporated a fragment of saquinavir into L-689,502. Bruce Dorsey, a new hire in 1989 in Vacca's group, and his associate, Rhonda Levin, succeeded in synthesizing indinavir. Although in the monotherapy trials around forty percent of the patients were below four hundred copies of RNA after six months on the drug, HIV developed resistance to indinavir in some patients. Fortunately, it was found that the combination of indinavir and AZT or lamivudine (Epivir) was quite effective in substantially suppressing the virus levels. Merck's studies of combination therapy were the first to prove the efficacy of the cocktail approach and became the standard for the industry. After filing with the FDA in January 1996, indinavir received approval in March 1996 in an accelerated review process.[32]

Other important early protease inhibitors included nelfinavir (Viracept, approved in March 1997) by Agouron (now Pfizer) and amprenavir (Agenerase, approved in April 1999) by Vertex. These are collectively known as the first-generation protease inhibitors.

Vertex was a bit late to the game of protease inhibitors. Heavily leveraging crystallography and computer-aided drug design, Vertex and their Japanese collaborators at Kissei Pharmaceuticals discovered amprenavir, a very potent, picomolar inhibitor of HIV protease. With the drug from Kissei, Vertex then collaborated with Burroughs Wellcome to investigate the PK properties of amprenavir in 1993. Drunken with the financial success of AZT, Burroughs Wellcome was late to the field of protease inhibitors and needed Kissei/Vertex's drug as a jumping board. Amprenavir was small enough to penetrate to the brain, where the virus often hid. Apparently liking what they saw, Burroughs Wellcome shut down their own small protease program and agreed to pay for the full cost of the $200 million for the clinical trials of the drug. Amprenavir was approved by the FDA in 1999 and Vertex and Burroughs Wellcome (which became later Glaxo Wellcome in 1995, and eventually Glaxo SmithKline [GSK] in 2000) jointly sold the drug with a brand name Agenerase. The drug was later discontinued for use because it had a poor aqueous solubility (0.04 g/mL) despite having a relative one hundred percent bioavailability. It required a high ratio of excipient-to-drug to afford gastrointestinal tract solubility and eventual absorption. Vertex proceeded to prepare its phosphate ester prodrug, fosamprenavir, which had an eight-fold increase of aqueous solubility (0.31 g/mL). The prodrug fosamprenavir was approved in 2005 and Vertex and GSK sold it under the brand name Lexiva.

Protease inhibitors are perhaps the most efficacious of the antiretroviral classes, in part because of their generally high genetic barrier to resistance. The early use of the first-generation protease inhibitors saquinavir, ritonavir, indinavir, and nelfinavir did not take advantage of boosting by CYP3A4 inhibition. Nevertheless, the suboptimal pharmacokinetic properties of these inhibitors meant that although the genetic barriers were high, resistance did ultimately emerge to these first-generation protease inhibitors. Later, a pharmacokinetic boost provided by ritonavir allowed for less frequent dosing, once-a-day for many drugs, and a higher genetic barrier to resistance, a consequence of the much-improved daily trough plasma concentrations.

2.3.3 Second-generation HIV protease inhibitors

Although the initial ritonavir-boosting regimens were implemented with the first-generation protease inhibitors, the development of more potent and safer second-generation inhibitors heralded a new era of cocktail regimens. These second-generation inhibitors include fosamprenavir (the prodrug of amprenavir), lopinavir, atazanavir, tipranavir, and darunavir.

Many of the first-generation protease inhibitors were peptomimetics. They were designed based on a modified peptomimetic motif wherein the scissile bond of a peptidic substrates is replaced by a non-cleavable transition-state mimetic. It invariably took a tremendous amount of medicinal chemistry to methodically modify the peptide in every minute detail to achieve a drug that was potent and, more importantly, bioavailable. Due to the vulnerability of the early protease inhibitors toward metabolism, nearly all of them had to be given with ritonavir as a PK booster. Upjohn took a different approach.

Upjohn in Kalamazoo, Michigan did not want to take the peptomimetic approach. Instead, they carried out a medium throughput screen on a dissimilarity set of five thousand of their historical compounds in the early 1990s. A library of five thousandcompounds seems absurdly small for a screen in light of today's high throughput screen (HTS) routinely going through millions of compounds. Nonetheless, they identified warfarin, a blood thinner, as a weak inhibitor of HIV protease. Coincidently, their neighbors at Parke–Davis in Ann Arbor, Michigan identified a similar substrate from related screening efforts. Even though warfarin was far less potent than peptomimitc protease inhibitors, it was still attractive because warfarin was vastly bioavailable. In essence, Upjohn favored pharmacokinetics over potency for their screening hits.

They subsequently carried out a focused screen of compounds similar to warfarin and identified phenprocoumon, a warfarin analog and a known orally active anticoagulant. Phenprocoumon was thirty-fold more potent than warfarin and had already began to show antiviral activities in a cellular assay. A crystal structure of the phenprocoumon–protease complex was very helpful to future drug designs of better protease inhibitors. Taking advantage of computer-aided, structure-based drug design, Upjohn arrived at their first-generation clinical candidate. But it was discontinued in favor of more potent analogs. Their second-generation clinical candidate was once again abandoned because it had a very high degree of protein binding and its potency was still modest, particularly in comparison to the most active contemporary peptomimetics. They eventually came up with their third-generation clinical candidate, tipranavir, with superior safety and efficacy. The drug, brand named Aptivus, was approved by the FDA in 2005. Regrettably, tipranavir seems to have more severe side effects compared to other protease inhibitors.[33]

The latest entrant of HIV protease inhibitors was Janssen's darunavir (Prezista), approved by the FDA in 2006. The discovery of darunavir by Arun Ghosh took a long and winding journey. After finishing his postdoctoral training with E. J. Corey at Harvard in 1988, Ghosh worked at Merck's West Point, Pennsylvania site for six years and was involved in the discovery of HIV protease inhibitors. But he decided to start his independent academic career at the University of Illinois at Chicago in 1994 before moving to Purdue University in 2005. To discover better HIV protease inhibitors, Ghosh employed a "backbone binding" design concept to maximize the interactions with the active site of HIV protease, particularly, to promote extensive hydrogen bonding with protein backbone atoms. His group initially installed a tetrahydrofuran (THF) ring onto Roche's saquinavir. Later, by combining the left-handed feature of Merck's indinavir, Ghosh arrived at a reasonably good THF-containing protease inhibitor. Meanwhile, Vertex/Kissei also incorporated the similar 3-(S)-THF motif as the P_2 ligand of their sulfonamide protease inhibitor. Interestingly, Searle actually had a priority on the sulfonamide-containing protease inhibitors and Vertex had to pay Searle $25 million in 1995 to resolve the patent issue even though Searle's own program flopped due to protein binding and extensive metabolism issues.

In the field of drug discovery, one always learns from the other. While Vertex "borrowed" Ghosh's THF fragment, Ghosh was not shy about "borrowing" Vertex's sulfonamide moiety. For Ghosh, to solidify his intellectual property position and boost potency via more hydrogen bonding, he invented a bis-THF group in place of the THF fragment. The result was d*arun*avir, whereas "arun" in the generic name was undoubtedly a nod to the inventor. Development of darunavir was carried out by Tibotec, a subsidiary of Janssen. The drug was approved in 2006 and Janssen sells it with a brand name of Prezista.[34]

The coverage of initial protease resistance by PK-boosted second-generation protease inhibitors allowed for the sequential use of protease inhibitors in patients who developed resistance to first generation agents. This is especially true for the more potent second-generation inhibitors.

At the end of the day, almost all HIV protease inhibitors, including the latest and the best, darunavir, need to be given with a pharmacokinetic enhancer such as ritonavir or cobicistat to achieve effective plasma drug levels at the desired dose and frequency. Gratifyingly, the availability of protease inhibitors has dropped the fatality rate for AIDS patients by seventy percent.

To summarize, ten HIV protease inhibitors currently on the market are:

- saquinavir (Invirase, Roche) in 1995,
- indinavir (Crixivan, Merck) in 1996,
- ritonavir (Norvir, Abbott) in 1996,
- nelfinavir (Viracept, Agouron/Eli Lilly) in 1997,
- amprenavir (Agenerase, Vertex/GSK, discontinued) in 1999–2008,
- lopinavir (Kaletra with ritonavir, Abbott) in 2000,
- atazanavir (Reyataz, Ciba Geigy) in 2003,
- fosamprenavir (a prodrug of amprenavir, Lexiva, GSK/Vertex) in 2005,
- tipranavir (Aptivus, Upjohn/Pfizer/Boehringer Ingelheim) in 2005, and
- darunavir (Prezista, Janssen) in 2006.

2.4 HIV Integrase Inhibitors

HIV is a retrovirus encoding fifteen proteins, of which only three have enzymatic activities: reverse transcriptase, protease, and integrase. HIV integrase helps integrate the viral DNA into host cellular DNA by catalyzing the integration of pro-viral DNA into the host cell DNA. This process is an essential step in the HIV viral lifecycle and the key step in establishing a permanent infection. HIV integrase is an attractive target because there is no cellular homologue in humans, thus integrase inhibitors offer selectivity and less chances of drug resistance. Because integrase enzyme catalyzes 3′-processing and strand transfer, joining the viral DNA to the host DNA. HIV integrase inhibitors are also known as HIV integrase strand transfer inhibitors (INSTIs).

Because there is no cellular protease homologue in humans, HIV integrase inhibitors would have less drug resistance. Even though inhibitors for HIV reverse transcriptase and protease were discovered soon after the discovery of the virus, it took more than twenty years for the drug industry to put the first HIV integrase inhibitor on the market. Merck's raltegravir was approved for marketing in 2007 as the first HIV integrase inhibitor, a culmination of considerable effort that was based on clearly defining the biochemical staging of enzyme function. One of the challenges was that the HIV integrase protein has a shallow and solvent-exposed binding surface. Early lead structures were frequently based on catechols, hydrazides, or coumarins, all of which failed to show antiviral activity in cell culture by a mechanism that could be reliably attributed to inhibition of virus genome integration.

The critical biochemical development was understanding the strand transfer step, in which the viral DNA was inserted into the host cell DNA. It was the key enzymatic process susceptible to inhibition rather than assembly of the enzyme on viral substrate or the 3′-cleavage reaction in which two nucleotides are removed from the termini of the double-stranded viral DNA. This mechanistic insight afforded a more effective screening assay. In 1999, using such an assay, Merck and a Japanese company Shionogi independently discovered diketoacid derivatives as the first specific inhibitor of HIV integrase that demonstrated antiviral activity in cell culture. These compounds bound to a complex of HIV and the viral

DNA substrate, with the diketoacid moiety, as a phosphate isostere, binding to the two magnesium divalent ions involved in catalysis, forming a ternary complex that interferes with the binding to host cell double-stranded DNA. By replacing the carboxylic acid with a tetrazole bioisostere, Shionogi was able to obtain the first inhibitor co-crystalized integrase. That was a great contribution to the field, even though the tetrazole analog did not become a drug due to stability issues.

Merck Research Laboratories in Rome, Italy succeeded in finding their integrase inhibitors by inter-breeding two drug discovery programs. In parallel to the HIV integrase program, Merck Rome also had a hepatitis C virus (HCV) program at the same time. A class of inhibitors of HCV NS5B RNA-dependent RNA polymerase (RdRp) had the dihydroxypyrimidine pharmacophore with a strong metal-binding capacity, even though the compounds *per se* did not inhibit HIV integration. Realizing that both HCV NS5B polymerase and HIV integrase rely on binding to the magnesium ion for their catalytic activity, Merck medicinal chemists used dihydroxypyrimidine as the more stable substitute for the diketoacid pharmacophore. Simple installation of a hydrophobic benzylamine gave rise to a very potent drug in an integrase strand transfer assay. Another major structural core change was methylation of the 2-nitrogen atom on the pyrimidine ring to convert the pharmacophore to hydroxypyrimidinone. Incremental modifications to improve physiochemical properties while maintaining cell penetration and limiting protein binding delivered a drug with an exceptional potency. That became raltegravir, which has been marketed as Isentress since 2008.[35]

Japan Tobacco discovered their HIV integrase inhibitor, elvitegravir, also from "inter-breeding" of two drug discovery programs. Their monoketo acid pharmacophore was derived from a quinolone antibiotic scaffold originally designed for bacterial DNA gyrase activity. Even at the very beginning, they had already found that simple 4-quinolone-3-carboxylic acid, but not the fancier 4-quinolone-3-glyoxlic acid, had some decent HIV integrase inhibitory activity. The rest of medicinal chemistry was more straightforward. After decorating the core structure that served as the key chelating ligand to bind the divalent magnesium ions, they arrived at elvitegravir. After the FDA approval in 2012, Japan Tobacco and Gilead co-marketed elvitegravir as Vitekta.[36]

GSK's integrase inhibitor, dolutegravir, employed a tricyclic carbamoyl-pyridone scaffold. It was approved in 2013 with trade name Trivicay. Five years later, GSK's "me-too" drug bictegravir, which was a very close cousin of dolutegravir, was approved in 2018. GSK sold it as a combination drug with emtricitabine and tenofovir alafenamide with trade name Biktarvy, which is now a constant feature of commercials on TV every day, everywhere.

In all, four integrase inhibitors in total are on the market:

- raltegravir (Isentress, Merck) in 2007,
- dolutegravir (Trivicay, GSK) in 2013,
- elvitegravir (Vitekta, Japan Tobacco/Gilead) in 2014, and
- bictegravir (Biktarvy with emtricitabine/tenofovir alafenamide, GSK) in 2018.

2.5 HIV Entry Inhibitors

HIV entry is one of the first steps for the virus to invade host cells, but HIV entry inhibitors were late in the game. So far only two HIV entry inhibitors are on the market. One is Roche's HIV fusion inhibitor enfuvirtide (Fuzeon), approved in 2003. The other is Pfizer's CCR5 antagonist maraviroc (Selzentry), approved in 2007. The two drugs block virus entry into the host cell at different points. Enfuvirtide prevents the fusion of the virus envelope with the host cell membrane by compromising the functions of gp41. On the other hand, maraviroc interferes with the interaction of the virus with one of its key co-receptors on the host cell.

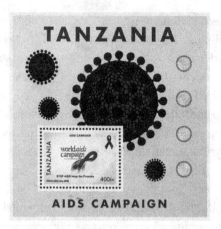

Fig. 2.5 AIDS Campaign © Tanzania
Post

2.5.1 HIV fusion inhibitors

When an HIV virion approaches the CD4 cell, fusion takes place
between the lipid membrane of the virus and the membrane
of the host cell. The HIV envelope glycoprotein is a polyprotein,
consisting of two no-covalently associated subunits: gp120 and
gp41, both of which are functional glycoproteins. This fusion is
triggered by the interactions of spike proteins on the surface of
the HIV envelope with specific cell surface receptors such as *CD4
receptor*, which is the main receptor for HIV to bind gp120. HIV
gp41 is deployed in a *trimeric* form on the virion surface in a met-
astable state that is maintained by its close association with gp120.
Inhibition of the fusion is a unique strategy to interrupt the HIV
lifecycle by preventing the HIV virion from making contacts with
host cell membrane. Ironically, the successful fusion inhibitor was
discovered from the failed attempts to produce AIDS vaccines.

Despite decades of intensive efforts to discover AIDS vaccines,
regrettably, none has materialized. But not all is lost, as enfuvirtide
was discovered during the process of finding an HIV vaccine.

Initially, synthetic peptides derived from gp41 were not targeted as inhibitors of HIV fusion. Rather they were investigated as part of epitope-mapping aimed at evaluating strategies in developing vaccines. Serendipitously, an antiviral effect was observed when those synthetic peptides were incubated with human T-cells. In 1992, scientists at Duke University Medical Centre prepared a synthetic peptide mimetic DP-107, a helical, thirty-eight-amino-acid molecule. It corresponded to an alpha-helix domain on gp160, which located at the N-terminal region of gp41. It was tested active to inhibit HIV replication. Subsequently, it was found that the C-terminus of gp41 also contained sequences that were predicted to form an alpha-helical structure. Later, DP-178, containing thirty-six amino-acids, was designed to correspond to another alpha-helix domain of gp160. The synthetic peptide DP-178 was tested to be one hundred times more potent than DP-107 in cell culture. A biotech company, Trimeris Inc., was founded to develop DP-178, which would gain the generic name of enfuvirtide. Roche bought the rights from Trimeris and carried out clinical trials which eventually led to the approval of the drug in 2003; its brand name is, fittingly, Fuzeon. Having a drug with a novel mechanism of action is beneficial, especially when drugs with other mechanisms failed or developed resistance. However, the peptidic nature of enfuvirtide prevents oral absorption, necessitating subcutaneous administration by injection twice daily. The inconvenient dosing regimen has relegated the drug primarily to use in salvage therapy for those with HIV infections whose viral load cannot be adequately controlled by oral medications.[37]

Small-molecule inhibitors that mimic the action of enfuvirtide have been sought since the elucidation of the mechanism of function of gp41 and the identification of a potential binding pocket for small molecules. But inhibitors of this type have largely proven to be elusive.

As alluded to earlier, HIV initiates access to its host cell by the binding of the viral envelope protein gp120, which is expressed as trimers on the surface of the virion, to the host T-cell receptor CD4, the prelude to a series of carefully choreographed events that offer multiple opportunities for therapeutic intervention. Small-molecule agents that interfere with the attachment of gp120 to CD4 have been described and are potentially attractive anti-HIV drugs. One such HIV attachment inhibitor, fostemsavir, the phosphonooxymethyl prodrug of BMS-626529 formulated as the tris(hydroxymethyl)aminomethane salt, is currently in clinical development. The prodrug element of fostemsavir is designed to increase the solubility of BMS-626529, which is designated as a BCS (Biopharmaceutics Classification System) Class 2 drug based on its poor solubility and high membrane permeability. The phosphonooxymethyl moiety increases the solubility of the compound in the gut, thereby reducing dissolution-limited bioavailability.

2.5.2 CCR5 inhibitors

CD4 alone, however, is not sufficient to permit HIV fusion and cell entry—an additional co-receptor from the chemokine family of G-protein coupled receptors (GPCRs) is required. The chemokine receptors CCR5 and CXCR4 have been demonstrated to be major co-receptors for the fusion and entry of HIV-1 into host cells. Approximately fifty percent of individuals are infected with strains that maintain their requirement for CCR5. Meanwhile, CCR5-deficient individuals are apparently fully immunocompetent, indicating that an absence of CCR5 function may not be detrimental and that a CCR5 antagonist should be well tolerated. Individuals homozygous for this deletion, which creates a nonfunctional receptor, are protected from contracting HIV infection. A mutation exists in about ten percent of Europeans and it helps to protect them from HIV infection. In a stunningly unethical approach, a Chinese biophysicist, Jiankui He, deleted the CCR5 gene from embryos using the CRISPR-Cas9 technique. He then implanted the genetically engineered embryos into a woman who subsequently gave birth to twin baby girls at the end of 2018. He assumed that deleting the CCR5 would prevent HIV from entering cells. Thus, deletion of CCR5 would make them less susceptible to HIV. For his unethical and criminal conducts, He was given a three-year prison sentence for illegally practicing medicine.

Schering-Plough was a pioneer in the field of CCR5 antagonists. From their high-throughput screening (HTS), they found two series of hits from their old muscarinic antagonists. From one series of piperidino-peperidines, they arrived at ancriviroc, which became the first CCR5 antagonist to advance to clinical studies. Although ancriviroc was safe and demonstrated a clear antiviral effect, the clinical development of ancriviroc was discontinued in 2005 due to the cardiac side effects (QT prolongation) that were noted at the highest dose tested. Binding to human *ether-a-go-go*-related gene (hERG) tightly was directly associated with QT prolongation thus hERG was an off-target associated with cardiotoxicities. Inhibition of hERG can result in potentially fatal arrhythmias associated with long QT syndrome, in which repolarization of the heart is slowed.

A high therapeutic index is particularly important in the HIV arena as drugs to treat HIV are rarely given in isolation, but rather in combination with other agents to prevent emergence of viral resistance. From another series of their HIT hits, piperidino-peperizines, Schering-Plough arrived at another CCR5 antagonist as clinical candidate, aplaviroc. But clinical trials for aplaviroc were also terminated when safety concerns surfaced about the occurrence of severe hepatotoxicity in several patients also in 2005. Meanwhile, the third CCR5 antagonist developed by Ono Pharmaceuticals and GSK was also discontinued in 2005. Schering-Plough brought the fourth CCR5 antagonist vicriviroc to clinical trials. But Merck, who acquired Schering Plough in 2009, decided not to proceed with a New Drug Application (NDA) because vicriviroc did not show robust efficacy. The graveyard of dead drugs was littered with failed CCR5 antagonists.

Pfizer's Sandwich site was the only one that successfully put a CCR5 antagonist on the market, launching maraviroc in 2007 with Selzentry as the trade name.

Pfizer's Sandwich obtained a piperidine derivative from their HTS campaign of 500,000 compounds in their corporate screening file. Their drug design focused on two issues. One was that CYP2D6 inhibition was a liability with early piperidine derivatives, which led to an approach that sought to sterically hinder the basic nitrogen atom as a means of interfering with its postulated association with the iron atom of the enzyme. This tactic culminated in the adoption of the tropane bicyclic ring system as the key scaffold for optimization. The second problem involved minimizing interaction with the cardiac potassium ion channel encoded by the hERG. Inhibition of the hERG ion channel was independent of the pK_a of the central basic amine, and both the 3-isopropyl-5-methyl-1,2,4-triazole and the difluorocyclohexylamide were identified as the key structural elements. After profiling over three thousand compounds, maraviroc emerged with excellent antiviral potency accompanied by modest hERG inhibition, a low molecular weight, and reasonable lipophilicity.[38]

Pfizer's Sandwich was one of the more productive sites at Pfizer, having brought a number of blockbuster drugs to market including Diflucan, Viagra, Norvasc, and Selzentry. Sadly, the site was closed in 2011 during the downturn of world's pharmaceutical industry.

HIV entry inhibitors on the market are:

- enfuvirtide (Fuzeon, Trimeris/Roche) in 2003, and
- maraviroc (Selzentry, Pfizer) in 2007.

2.6 The Road to Eradication

Following the approval of AZT in 1987, dozens of additional direct-acting antiretroviral drugs have been approved. They belong to six different classes that includes eight nucleoside (nucleotide) reverse transcriptase inhibitors, seven nonnucleoside reverse transcriptase inhibitors, ten discrete protease inhibitors, four integrase inhibitors, and two HIV entry inhibitors. These drugs target the three essential enzymes encoded by the virus: reverse transcriptase, protease, and integrase. When used in combination, these drugs represent an effective therapeutic regimen to control the replication of HIV, reducing the disease to a manageable, chronic infection, but they are not curative. Missed drug treatments contribute to the development of drug resistance due to the archived nature of the virus within long-lived memory T-cells, necessitating treatment courses that are life-long.

Impressively, nearly one-half of all antiviral drugs are used for the treatments of AIDS. The other class of major antiviral drugs are anti-HCV drugs. The lessons that we learned from the

Fig. 2.6 AIDS © Serbia Post

pharmacology of HIV and AIDS greatly aided the work on drugs to treat HCV and SARS2-CoV-2.

Long-acting HIV/AIDS drugs are in the frontier of the field. In 2020, Canadian authorities approved Marketing of ViiV's Cabenuva, a long-acting HIV drug comprised of cabotegravir, an integrase strand transfer inhibitor, and rilpivirine, a non-nucleoside reverse transcriptase inhibitor. It only needs to be given once a month as an injection. Doubtlessly, Cabenuva will help tremendously with patient compliance.

Drug therapy for HIV disease has been remarkably successful at extending life, with mortality rates of those on cocktail regimens approaching that of the general population.

3

Hepatitis Viruses

When you have eliminated the impossible, whatever remains, however improbable, must be the truth.
—Arthur Conan Doyle (1859–1930)

Hepatitis is the inflammation of the liver and the development of jaundice is a hallmark of hepatitis. Symptoms of jaundice include vivid yellow color to the whites of the eyes, sometimes yellow skin of entire body, exhaustion, nausea without vomiting, urine of the color of tea, and putty-colored feces. Diseases resembling hepatitis have been known since antiquity. There were documents describing hepatitis-like yellow jaundice outbreaks in China five thousand years ago. In the fifth century, Hippocrates (460–375 BC) recorded a benign epidemic jaundice, which was likely caused by a hepatitis virus. He termed the skin-yellowing condition *ikterus*. In history, hepatitis epidemics assailed towns and crippled troops over the ages following closely with wars. In nineteenth-century Europe, German soldiers called hepatitis *kriegsikterus* (war jaundice) and the French *grande armée* dubbed it *jaunisse des camps* (camp jaundice). During the American Civil War, more than forty thousand cases of hepatitis erupted in the Union troops. World War II saw both the Allies and the Axis encountering viral hepatitis as a major medical issue. In 1942, there was a major outbreak of hepatitis in the US Navy when fifty-six thousand sailors were infected following administration of the yellow fever vaccines stabilized with normal human serum, which was likely contaminated with the hepatitis

Conquest of Invisible Enemies. Jie Jack Li, Oxford University Press. © Oxford University Press 2022.
DOI: 10.1093/oso/9780197609859.003.0003

Fig. 3.1 National Viral Hepatitis
Control Program © India Post

virus. One year later, viral hepatitis spread quickly in troops in
North Africa and the Mediterranean. In the Mediterranean theatre
alone, more than thirty thousand cases of jaundice were reported in
a two-year period.[1]

Hepatitis may be caused by five types of viruses, including hepatitis A, B, C, D, and E viruses, although there may be more subtypes
to be discovered. The discovery of the hepatitis viruses is one of the
most fascinating scientific adventures in the last sixty years, which
in turn, has resulted in effective serological tests, vaccines, and effective therapeutics. Most impressively, direct-acting antiviral
drugs for the treatment of hepatitis C are so efficacious that most
patients can now achieve a sustained virologic response (SVR),
which is very close to a cure. It is not over-reaching to predict the
eradication of hepatitis C via successful vaccination, similar to
polio and measles.

We begin our story with the most prevalent hepatitis A virus.

3.1 Hepatitis A Virus

Hepatitis A was known as traveler's hepatitis and epidemic hepatitis in the general public and, more popularly, as *infectious hepatitis*

among virologists. Hepatitis A virus (HAV) was not discovered until 1973 despite having ravaged humanity for millennia. Oddly enough, HAV was discovered eight years *after* Baruch Blumberg discovered the hepatitis B virus (HBV) in 1965.

3.1.1 Discovery of the hepatitis A virus

The silver lining of the viral hepatitis outbreaks during WWII was that they stimulated intensive research in the field during and after the war when virology entered a golden age. It was soon recognized that there were two forms of hepatitis. One form was highly contagious with a shorter incubation period of about one month, which occurred in epidemic form in areas of poor hygiene and sanitization. Contaminated food and water were common sources of infection. It appeared to be transmissible by the fecal–oral route: the virus replicated in the liver, secreted into bile, and was then shed in the stool. The other form was less common, with a longer incubation period of two to three month, and was spread by inoculation of infected blood, serum, or plasma. In 1947, British liver specialist Frederick MacCallum suggested that the two forms of hepatitis should be called hepatitis A and B, respectively.

Between 1956 to 1965, Saul Krugman, nephew of Albert Sabin and a Rockefeller Institute-trained virologist at the New York University, and his colleagues conducted hepatitis studies at the Willowbrook State School for intellectually handicapped children on Staten Island, New York. They intentionally infected those children with hepatitis viruses. The ethics of their conduct was quite controversial even in that era, although their investigation provided a definitive answer to MacCallum's proposal. Meanwhile, the US Army obtained large quantities of the infectious hepatitis inoculum by infecting "volunteer" inmates with hepatitis at the federal prison in Joliet, Illinois. Their experiments with those sera further confirmed that indeed two distinct forms of viral hepatitis existed, just as MacCallum suggested. Today, the practice of using prison inmates for clinical trials is now deemed unethical and is no longer in use.

An animal model for hepatitis A was established in 1967 when it was reported that human hepatitis virus could be passaged in marmoset monkeys. Five years later, the HAV was identified in 1973 at the National Institute of Health (NIH) using immune electron microscope and specific serology to detect the HAV antigens.

Stephen Feinstone joined the NIH in 1971 immediately after finishing medical school to avoid the draft to the Vietnam War, like many of his peers. That trend injected many talents to the NIH at the time, known as "Yellow Berets," and unintentionally made the NIH the powerhouse of medical research. To enter the program during the "doctor draft" era at the NIH was extremely competitive, and the work performed produced a total of nine Nobel laureates as an unintended consequence of avoiding the Vietnam draft.

Feinstone was assigned to the Viral Hepatitis Group, which was headed by a prominent virologist, Robert Purcell, who made many significant contributions to hepatitis virus research. Purcell directed Feinstone and another research associate to find the virus responsible for hepatitis A in stools. After the first stool extraction, the other research associate decided he was not interested in hepatitis A after all; or was it just the smell?! As a consequence, Feinstone was alone to do the stool extractions, which did not endear him with his colleagues. In fact, the odor produced from his extractions pervaded the entire building and everyone knew when it was a stool extraction day! In October 1973, Feinstone successfully observed hepatitis A virus as antibody-covered particles under an immune electron microscope, recently developed by his coworker Albert Kapikian. Purcell, Feinstone, and Kapikian subsequently proved the specificity of the particles using a serological antibody assay. They decided to submit the paper on their discovery to the *Science* magazine, published by the American Association for the Advancement of Science (AAAS) in downtown Washington, not far from the NIH campus in Bethesda, Maryland. The trio was able to submit their manuscript in person to the *Science* magazine office and the paper was published three weeks later. Their landmark paper would have made the headline on the *Washington Post* if the Watergate scandal had not surfaced on the same day.[2]

HAV is a non-enveloped, spherical, positive-strand RNA virus. The virion of HAV is "naked," without a protective lipid envelope that is common in many other viruses. It is among the smallest and structurally simplest of animal RNA viruses.

In 1988, a hepatitis A epidemic broke out in Shanghai, China due to the ingestion of raw shellfish, mostly HAV-contaminated clams. There were approximately 310,000 cases of hepatitis A and over eight thousand patients required hospitalization. At the end of the outbreak, forty-seven patients had died, giving rise to a mortality rate of 0.015 percent. The mortality rate for patients who already had hepatitis B was 5.6 times greater than patients who contracted HAV alone.[3]

3.1.2 Hepatitis A vaccines

There is no effective antiviral drug treatment available for the hepatitis A disease today. Good hygiene is probably the best preventive measure. Although a vaccine would have been ideal as a prophylactic, it did not materialize until 1991, when the first inactivated HAV vaccine, SmithKline Beecham Biologicals' Havrix, was licensed in Europe. In America, Havrix was approved in 1995 by the FDA and Merck's HAV vaccine Vaqta was also approved in the same year.

Before the emergence of an HAV vaccine, the only prophylactic agent available was passive immunization with serum immunoglobin (Ig). Immunoglobin is an old term for today's antibody, which is a Y-shaped protein produced by B-cells of the immune system in response to exposure to antigens. It was basically concentrated antibody prepared from pooled plasma from convulsant patients who already had hepatitis A. Not only was it expensive and difficult to make, it had only a short period of protection—merely three months. Therefore, serum immunoglobin was largely used only for babies. In contrast, vaccines are vastly superior to serum immunoglobin and HAV vaccines provide a robust protection against the virus infection for over twenty years.

To develop a vaccine against the virus, one often has to first be able to grow the virus in cells. The golden era of vaccines began in 1949 with the discovery of virus propagation in cell cultures when John Enders showed that polio virus multiplied in cultures of monkey kidney cells. Not only did the far-reaching discovery earn Enders the 1954 Nobel Prize, but it also made it possible, for the first time, to make the polio vaccine. More important, it showed the way in which many other viruses might be grown in cell lines.

Immediately after the discovery of HAV, many attempts were made to grow and isolate the virus in cells without much success. In the early 1960s, an HAV cell line called Detroit 6 turned out to be a fluke.At the time, scientists at the Parke–Davis Company in Detroit reported a cell line derived from human bone marrow. Named the Detroit-6, they reported that they had isolated the etiological agent of hepatitis A in Detroit-6 cells and could recognize its presence by the characteristic *cytopathic effect*, a host cell's structural changes from viral infection. But the data was hard to confirm and Parke–Davis wisely abandoned the cell line. But Ruth Cole, the technician who took part in the Parke–Davis studies, went back to her native Australia after she got married. Her husband secured a research fund for her, and she continued research on Detroit-6 cells and published papers in prestigious journals. Suspecting the data was too good to be true, a few of her colleagues presented her with coded samples to test. But Cole failed to show that the observed cytopathic effects were specific to hepatitis A. Was it a scientific fraud in this case or simply self-deception on Cole's part? The answer is irrelevant now because cytopathic effects could be very subjective.

It was later suggested that the cytopathic effects were most likely caused by a parvovirus contaminant.[4]

In 1967, Freidrich Deinhardt in Chicago showed that the HAV could be propagated in marmoset monkeys. In 1979, Philip Provost and Maurice Hilleman at Merck eventually succeeded in growing HAV in the CR326 cell line. Using HAV-infected marmosets, they recovered the CR326 strain of human HAV and then repeatedly propagated the virus in primary explant cell cultures of marmoset livers and in the normal fetal rhesus kidney cell line. The ability to grow live virus *in vitro* was very significant for the production of HAV and for making the vaccine. The *New England Journal of Medicine* touted the achievement "a giant step forward" in the search of a vaccine.[5] After purification of the HAV particles, Merck initially tried the attenuated-virus vaccine approach with some success; however, they later decided to make the killed-virus by inactivating the virus with formalin (eight percent formaldehyde aqueous solution). The deactivated virus particles were then absorbed into an aluminum hydroxide adjuvant to enhance the vaccine's immunogenicity. After testing the vaccine with success on marmoset monkeys, clinical trials revealed that the vaccine was safe and immunogenic in healthy human subjects. Three doses were required to evoke an adequate antibody response and to provide full protection against the HAV. Merck's HAV vaccine, Vaqta, was sold with a price tag of $50–60 per dose at the time. SmithKline Beecham used a different virus strain, HM175 to prepare Havrix with 2-phenoxyethanol as a preservative. Both Havrix and Vaqta had a remarkable efficacy of ninety-seven percent and one hundred percent, respectively.[6]

Efficacious HAV vaccines containing a formalin-inactivated virus produced in cell culture have now been licensed in multiple countries. In the United States, HAV infection has declined substantially since 1995, when vaccinations were recommended for individuals at risk. Today, the most likely means of contracting hepatitis A for Americans is traveling to South American countries and some other developing countries. This is why people traveling to these locations are recommended to get vaccinated for hepatitis A.

The People's Republic of China took a different approach toward the HAV vaccines. They used the H2-live vaccine. The H2 viruses were derived and isolated from fecal specimens of hepatitis A patients. They were then propagated in human fibroblast, attenuated by serial passages (occasionally as many as 23 times!) from monkey kidney cells to human fibroblasts. It was followed by purification from cell lysates, formalin inactivation, and absorption to aluminum hydroxide adjuvant. The attenuated H2 vaccine was introduced in China in 1992, followed by the introduction of inactivated HM-vaccine in 1993. Millions of Chinese citizens have been vaccinated. In China, HAV elimination with HAV vaccines is now a strong possibility.[7]

3.2 Hepatitis B Virus

Before the identification of the HBV, hepatis B was often known as the *serum hepatitis* as opposed to *infective hepatitis*, which referred to hepatis A. Make no mistake—hepatitis B is also highly contagious, sometimes acutely, even though it often only manifests as a mild illness. Populations more prone to contract hepatitis B include drug addicts who share needles, hemophiliacs who take transfusions of factor VIII, and people who engage in anal sex. During WWII, in 1942, the American military inoculated three hundred thousand men with a yellow fever vaccine in anticipation of a fight in the tropic Pacific region. Unfortunately, the vaccine was contaminated with HBV. As a result, fifty thousand men were hospitalized and more than one hundred died. Believe it or not, the battle of Midway was won by air pilots, some of whom were suffering from jaundice. Meanwhile in North Africa, an American tank commander laid low with hepatitis B fought to a draw with

Erwin Rommel's division of the German army, many of whom were sick with hepatitis A.

It is estimated that 400 million people worldwide have been infected with HBV while more than 350 million of them become chronic carriers. Chronic HBV infection leads to the development of cirrhosis (liver scarring), decompensation, and hepatocellular carcinoma, which accounts for over five hundred thousand deaths per year worldwide. In the United States, twelve million people have been infected with HBV at some time in their lives. Of those individuals, more than one million people have subsequently developed chronic hepatitis B infection. These chronically infected persons are at highest risk of death from cirrhosis and liver cancer. In fact, more than five thousand Americans die from hepatitis B-related liver complications each year. In many Asian and African countries, where the HBV is endemic, up to twenty percent of the population may be carriers. Transmission occurs primarily through perinatal (the virus is passed to babies from nursing mothers) or early childhood infection. In some of these areas, the perinatal transmission rate may be as high as ninety percent!

3.2.1 Discovery of the hepatitis B virus

The saga involving the discovery of hepatitis B virus, like many scientific ventures, was a blend of genius thinking, teamwork, rivalry, and serendipity.[8]

Baruch (Barry) S. Blumberg, a grandchild of Eastern European Jewish immigrants, was born on July 28, 1925 in Brooklyn, New York. His father was a lawyer, providing the family a middle-class lifestyle. Blumberg was initially educated in an orthodox Jewish parochial school but moved to a more mainstream educational track, finishing up his undergraduate education at Union College in upstate New York. After earning his MD in 1951 at Columbia University and an internship at Bellevue Hospital in

1955, Blumberg studied at Oxford University, investigating polymorphism, which is the occurrence of two or more different morphs or forms in biology. After obtaining his PhD in 1957, he started his first real job at the NIH to study inherited polymorphism of proteins in different ethnic groups. Even though his supervisor, DeWitt "Hans" Stetten, did not understand or appreciate his esoteric and obscure research, the NIH kept funding Blumberg's research in case he knew something that they did not. Working with his friend at Oxford, Anthony Allison, Blumberg discovered a new polymorphism in 1960. They showed that a low-density cholesterol-transporting protein, called beta-lipoproteins, existed in varied forms in different populations as a result of genetic polymorphism. They made the discovery using the Ouchterlony double imminodiffusion test. A noted authority on immunology, Professor Örjan Ouchterlony (1914–2004), at the University of Gothenburg in western Sweden, introduced the test back in the 1940s. Also known as *passive double immunodiffusion*, the test is an immunological technique to detect, identify, and quantify antigens and antibodies. For Blumberg, the discovery that the blood of transfused subjects reacted with uncommon antigens became an important tool for his research.

The field of hepatitis virology was rapidly maturing in the early 1960s. Krugmen already confirmed MacCallum's proposal of the existence of hepatis A and B, respectively. In 1962, Frank MacFarlane Burnet, an eminent virologist from Australia and Nobel laureate in 1960, stated: "I think the greatest intellectual prize that a virologist can hope for is that someday he should be the first to explain the real natural history of serum hepatitis." Little did he know that day was merely a few years away.

In 1963, Harvey Alter, a young physician working at the NIH's Blood Bank, became interested in Blumberg's research in polymorphism and seconded in his laboratory. Alter would, in 2020, nearly sixty years later, win the Nobel Prize for his discovery of the hepatitis C virus (HCV). Longevity is good for Nobel laureates, just in case the prizes do not come in their early careers, as for both Peyton Rous and Harvey Alter.

In one of Alter's experiments on lipoproteins using the Ouchterlony test, he saw a precipitin reaction that was different from the typical lipoprotein bands. The antibody was isolated from a twelve-year old hemophiliac boy who had undergone many blood transfusions. While quite different from the low-density lipoproteins that they often encountered, it reacted with the serum from an Australian aborigine provided by a professor from Australia. The enigmatic serum protein antigen was named the *Australia antigen*, which was the *hepatitis B surface antigen* as we now know today. Others might have dismissed this obscure finding as an irrelevant curiosity because it was very rare. But Blumberg's hypothesis-generating mind was set in motion.

In 1964, Blumberg joined the Fox Chase Cancer Center in Philadelphia as an associate director to continue the same line of research. Not only did he become a big fish in a small pond, but he was also given freedom in both funding and personnel. The freedom was invaluable to the discovery of HBV because he did not have to justify the directions of his research to grant review committees—a freedom rarely granted nowadays!

The term *antigen*, which stands for *anti*body *gen*erator, warrants some explanation here. Antigens, often foreign proteins, are substances that can induce an immune response in the body. For instance, both bacteria and viruses may be considered antigens. Even food and pollens could be antigens to cause allergy. When encountering an antigen, the human body secrets antibodies and the epitopes of an antigen bind to the paratopes of the Y-shaped antibody protein, resulting in the *antigen–antibody reaction*. The antigen–antibody reaction was first recognized by Austrian scientist Karl Landsteiner (see Figure 3.2). In the 1900s, Landsteiner introduced the *typing* of red blood cells using different antigens. The four types of blood groups according to their red blood antigens are A, B, O, and AB.

Fig. 3.2 Karl Landsteiner © Guyana
Post

For blood transfusion, when type A blood is transfused to a patient with type A blood, there is no antigen–antibody reaction,
just like a patient's serum does not react against her own blood
cells. However, for a type A patient receiving type B blood, the
antigen–antibody reaction takes place when the patient's naturally occurring antibody in the plasma reacts with type A antigen
on type B red blood cells. The consequence is *agglutination*, which
causes red blood cells to clump together or lysing of the cells; that
is, a disintegrating of the cell and a leaking of their content of the
pigment and *hemoglobin*. Agglutination makes transfusion dan-
gerous, even lethal sometimes. Here, the antigen is an *agglutinogen*.
Ironically, Landsteiner published his seminal discovery about blood
types in 1900 as a footnote in a paper on pathologic anatomy in
which he described the agglutination that might occur when blood
of one person was brought into contact with another. Apparently,
he did not recognize the importance of the blood groups, possibly
because it was before the blood coagulation problem was solved
by adding sodium citrate as an effective anticoagulant. Blood
transfusions and transplantations literally saved millions of lives
and Landsteiner was deservedly awarded the Nobel Prize in 1930
for his "recognition and discovery of the human blood groups."

At Fox Chase, Blumberg observed under electron microscopy that serum containing the Australia antigens had millions

of virus-like particles. The particles had two types: large particles and small particles. The large particles (~forty-two nm in size) resembled complete viruses, which were indeed the complete virus because they could transmit hepatitis. The small particles (~twenty-two nm in size) were actually the Australia antigens because the small particles alone did not transmit hepatitis and they were later demonstrated to be protein coating fragments of the virus. The Australia antigens vastly outnumbered the active virus particles by roughly 1000:1. A few years later, in 1969, Blumberg astutely reasoned that the small, non-infectious particles might lead to immune response and thus provide the basis of a vaccine. He proposed that a vaccine for HBV could be prepared from the plasma of HBV-infected patients, unlike all previous vaccines that were prepared from animal tissue cultures (see next section, 3.2.2. Hepatitis B vaccines).

From 1964 to 1965, Blumberg and his colleagues observed near total absence of the Australia antigen from the sera of normal American subjects. But the antigen was common in the sera of people from Taiwan, Vietnam, Korea, the Central Pacific, and Australian aborigines. Most of the sera containing Australia antigens were from patients who had blood transfusions, which gave them an inkling that they might be associated to a blood-transmitted infectious disease. They also discovered the association between the Australia antigen and patients inflicted with Down syndrome who also had leukemia. (Down syndrome patients are at high risk of developing leukemia.) They initially proposed, erroneously, the possibility of an association between the Australia antigen and leukemia. During their ensuing investigations, random incidents began to provide the Eureka moments necessary to pinpoint an association of the Australia antigen with viral hepatitis. In 1967, Barbara Werner, a technician in Blumberg's laboratory, suddenly tested positive for Australia antigen even though she had been routinely using her and other colleagues' blood samples as the negative controls. She was the one who carried out the test of her blood sample herself! It was obvious that she contracted hepatitis, most likely from her handling of HBV-containing sera. In a very similar case, a Down syndrome patient suddenly tested positive although he had already tested negative several times for the Australia antigen. It was found that he had a mild case of hepatitis, offering a strong clue that the Australia antigen and hepatitis were related. Meanwhile, from screening several hundred sera from Claxton, a rural community in Georgia, the only positive sample with the Australia antigen was a hepatitis patient. To fulfill Koch's postulates, Blumberg tried very hard to grow the virus in tissue or organ cultures without success. He even hired, from the famous Werner and Gertrude Henle group, a virus specialist, who also failed to grow the virus despite three years of attempts.

With evidence mounting, Blumberg wrote a paper linking the Australia antigen to the hepatitis B virus in 1967. The manuscript was soundly rejected because many similar claims had been resoundingly proven incorrect. Eventually, the paper was published in the *Annals of Internal Medicine*. Inspired by Blumberg's exploits, Alfred Prince at the New York Blood Center discovered a *serum hepatitis antigen* in the blood of hepatitis B patients in 1968. It soon became evident that Prince's serum hepatitis antigen and Blumberg's Australia antigen were exactly the same protein. Prince's work strongly supported the notion that the virus Blumberg had identified was analogous to the post-transfusion hepatitis virus that had long been inferred from the earlier research in the field. Subsequently, one-by-one, groups around the world confirmed that the Australia antigen was indeed a marker for both acute and chronic hepatitis B and there were apparently healthy Australia antigen carriers.[9]

In 1970, David Dance and colleagues at the Middlesex Hospital in London reported visualizing, using electron microscopy, the whole virus particle (forty-two nm in size) that included nucleic acid. The whole HBV particles are now known eponymously as the *Dane particles*.

Because Blumberg and Prince solidly established the relationship between the Australia antigen and HBV, a serological test was readily worked out. Blumberg promoted the serological test using the immunodiffusion assay to identify HBV carriers. Testing surface antigen carriers was tremendously useful for blood banks and greatly reduced the incidents of HBV contamination. In 1972, Abbott Laboratories in North Chicago developed a new testing instrument named Austria-125 for highly sensitive detection of antigens or antibodies using the solid-phase sandwich radioimmunoassay. They used iodine-125-labeled antigens and such antigens bound to antibodies could be detected with high sensitivity.

To Blumberg's surprise and chagrin, the leading researchers—the mavens or the high priests of the discipline as he called them—were rankled. They were jealous because Blumberg discovered HBV while looking at quite different things. He made a claim for the discovery of the virus that others had sought for so many years but failed. But Blumberg's work linking the Australia antigen and the surface antigen of HBV was a landmark achievement. The Nobel Prize in 1976 was bequeathed to Blumberg and Carleton Gajdusek "for their discoveries concerning new mechanisms for the origin and dissemination of infectious diseases." Some in the scientific community believed that Alfred Prince should have shared the Nobel Prize with Blumberg.

After receiving his Nobel Prize, Blumberg visited China, India, and Africa to investigate liver cancer and promote HBV vaccines. It is remarkable that his pursuit of a relatively obscure subject led to unexpected findings that impacted millions of people's lives. To honor his momentous contributions in the field of hepatitis, the World Health Organization (WHO) in 2010 chose Blumberg's birthday, July 28, as World Hepatitis Day. Sadly, eighty-five-year-old Blumberg passed away in 2011 due to a heart attack (see Figure 3.3).

While HAV is an RNA virus, HBV is a small, enveloped, *DNA retrovirus* that exists with eight genotypes. HBV belongs to the

Fig. 3.3 Baruch Blumberg
© Angola Post

Hepadnaviridae family and causes hepatitis in both humans and animals. The HBV genome comprises a relaxed, circular, partially double-stranded DNA of approximately 3,200 base pairs that encodes the large, middle, and small surface proteins, core and precore proteins, DNA polymerase, reverse transcriptase enzyme, and the X protein. After entering hepatocytes, the relaxed circular DNA is converted to a covalent closed circular DNA, which functions as the transcriptional template for the four viral RNAs that are translated to the HBV proteins (see Figure 3.4).

3.2.2 Hepatitis B vaccines

In early 1969, anticipating a federal funding cut, scientists at the Fox Chase Cancer Center were encouraged to generate revenues from their research. Thus, patenting their research was strongly encouraged. Although having no experience on vaccines, Blumberg showed great insights when he formulated an invention of an HBV

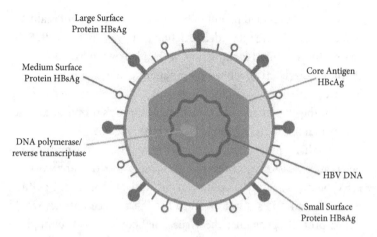

Fig. 3.4 The structure of the hepatitis B virus, diagram by
Alexandra H. Li

vaccine. Blumberg and Irving Millman recognized that a vaccine
for HBV could be prepared from the Australia antigen particles in
plasma, unlike all previous vaccines that were prepared from animal
tissue cultures. They observed that the virus produced very large
quantities of the small non-infectious particles containing only the
surface antigen. This is probably an immunologic strategy of the
virus, producing an excess amount of the antigen in the blood to
divert the antibodies produced by the host's immune system to de-
vour surface antigen. Therefore, the whole virus is spared. In their
patent, Blumberg and Millman proposed to separate the surface antigen particles from larger infectious particles using centrifugation
in a medium, such as a sugar or cesium chloride solution. Enzymes
were then added to remove any serum proteins remaining. Any
viable virus that might remain was then impaired or destroyed.
Killing the viruses with formalin was followed by column separa-
tion. Eventually, substances to increase the antigenicity of the surface antigen were used to increase the ability of the vaccine to elicit
a protective response from the vaccinated person.

Even with an issued patent in hand, Fox Chase was not really the place for translating their idea to actuality. It was not until 1975, six years after the initial formulation of the idea, when Merck in West Point, Pennsylvania licensed their patent. Under the leadership of Maurice Hilleman, Merck's Virus and Cell Biology Department took on the challenge of transforming Blumber's proposal to practice by making the first plasma-based HBV vaccine.

Merck had enjoyed great successes of making polio, mumps, rubella, and measles vaccines. But their old tricks did not work for HBV because the virus could not be cultivated in a cell culture. Because antibodies developed against the surface antigen would afford protection against the virus, they could obtain enough of the surface antigen from the blood plasma of persons who were carriers, just as Blumberg predicted. After purification of the virus, Merck came up with two methods for inactivating the virus, one physiochemical and the other entirely chemical. Hilleman and his team chose the chemical approach that applied three virus-killing procedures, one of which was the time-tested formalin treatment. That approach yielded antigen that was nearly one hundred percent pure. Krugman, despite his questionable bioethics for what he did with the intellectually disabled children at the Willowbrook State School, volunteered to be the first guinea pig. In 1975, Krugman injected the first Merck HBV vaccine into himself, his wife, and nine Merck supervisory personnel. In 1978, clinical trials were carried out among homosexual males who were most prone to the HBV infection. After a thirteen-year-long pioneering program, the vaccine was licensed to market under the apt trade name *Heptavax-B*, the first *subunit viral vaccine* produced in the United States. This was such a unique method of making a vaccine that none had been prepared in this manner before, and none have been since.[10]

With the success came some headaches. For one thing, the complex manufacturing process was too long, taking sixty-five weeks, the longest production time of any vaccine being produced. For another thing, it was simply too difficult to locate an adequate supply of suitable carrier plasma, which was obtained from donors chronically infected with hepatitis B, who were mostly homosexuals and intravenous drug users. On top of that, the procedures also endangered plant workers isolating the antigen from the plasma, which contained a live virus that could infect them. Last, but not the least, the vaccine was quite expensive, $100 for three doses needed. The Merck team therefore turned their attention to the Alexander cells, which were a line of liver cancer cells derived from an HBV carrier whose last name was Alexander. Even though they succeeded in developing a vaccine from Alexander cells as efficacious as *Heptavax-B*, Merck decided to forego this approach as well. Instead, they opted to make their next generation of HBV vaccine by employing the latest recombinant DNA technology.

In the early 1970s, cloning and genetic engineering applying the recombinant DNA technology began to achieve many technical breakthroughs. One novel technique, called "genetic splicing," enabled inserting new genetic information into a plasmid, or a circular DNA. Plasmids are self-replicating episomal units that are present in microbial cells. Merck initially succeeded in expressing hepatitis B surface antigen in the bacterium *E. coli* from their collaboration with Professor Bill Rutter at the University of California at San Francisco (UCSF). But the bacterium produced a form of the antigen that was slightly different from the natural one—the two proteins folded differently. Merck then turned to Professor Ben Hall, a leading yeast geneticist at the University of Washington at Seattle. Hall inserted the antigen DNA to ordinary bakers' yeast, which produced an active surface antigen that in turn evoked immunogenicity similar to that of plasma-based *Heptavax-B*. For manufacturing, Merck collaborated with Chiron Corporation, a biotech company founded by Hall in Emeryville, California. Chiron had provided proof of the principle that the antigen could be made in yeast and provided the vectors, which helped Merck to jumpstart their recombinant vaccine. A vector, such as a plasmid or a virus vector, is a DNA molecule that carries foreign genetic material into another cell where it can be replicated. A vector containing foreign DNA is termed a recombinant DNA. After inserting the genetic information into the yeast plasmids, fermentation of the bioengineered bakers' yeast was done in large fermentation vats. In due time, the yeast was then separated from the fermentation broth by centrifugation. Rupture of the yeast cells released the antigen, which was subsequently separated, ultra-filtered, and purified. After inactivating the antigens with formalin, the purified surface antigens were adsorbed to aluminum hydroxide adjuvant to boost the vaccine's immunogenicity. The vaccine was then preserved with thimerosal, a mercury-containing organic compound that has been the center of some controversies during the last few decades. Ill-informed anti-vaxxers erroneously blamed thimerosal for causing autism in children, a notion that has long been debunked.[11]

Clinical trials of the recombinant DNA vaccine showed that eighty to one hundred percent of the volunteers had produced antibodies specific to the hepatitis B surface antigen with no serious adverse effects. This was on par with *Heptavax-B*. This first human vaccine prepared by cloning was approved in 1986 and was sold under the trade name, appropriately, *Recombivax HB*. It was also the first licensed vaccine to prevent human cancer because chronic hepatitis B can cause hepatocellular carcinoma. In retrospect, HBV is luckily a distinctive antigen that spontaneously forms highly immunogenic virus-like particles and is inherently a potent immunogen. Not all antigens are as highly immunogenic as that of HBV.

Working with the WHO, Merck strived to bring *Recombivax HB* to the world. Unfortunately, unlike the significantly cheaper ivermectin, *Recombivax HB* was too expensive ($180 for a course) to give for free liberally. In 1987, Merck famously donated ivermectin (Mectizan) free to anyone in the world who needed it. Mectizan is an anti-parasite drug that is very efficient in preventing the river blindness, a black fly transmitted disease that plagued sub-Sahara Africa for decades.

Like many developing countries in Asia, China was ravaged by hepatitis B infection. With a fifteen percent infection rate in such a mammoth country, hepatitis B was the number one public health problem. After Nixon visited China in 1972, Blumberg was eager to visit China and promote HBV testing and, later, the vaccine. But the invitation was not extended to him until he won the Nobel Prize in 1976, which coincided with the fall of the end of the devastating Great Culture Revolution. Blumberg traveled to China in 1977 and transferred knowledge of his team's research and later set up contact with Merck. At the end of 1980s, the Chinese Government approached Merck about the HBV vaccine. It was immediately realized that Merck's genetically engineered HBV vaccine *Recombivax HB* was too expensive; therefore, technology transfer was the only financially viable approach. Merck CEO Roy Vegelos decided to charge the Chinese government a rock-bottom price of $7 million for two sets of state-of-the-art recombinant DNA vaccine plants capable of producing twenty million doses a year. One set was built in Beijing and the other in Shenzhen. Soon, with universal vaccination, China was able to bring the scourge of hepatitis B under control and millions of lives were saved.[12] Even today, Chinese people are still grateful to Merck, in general, and Dr. Vegelos, in particular. Chinese people never forget friends in need. For instance, Canadian surgeon Norman Bethune went to China to support the Chinese fight against Japanese invaders during WWII. He perished due to a bacterial infection during a surgery just before penicillin became available. The Chinese people were so grateful to Dr. Bethune that there was a Norman Bethune Hospital in every major city in China in the 1950s.

In Asia, the prevalence of HBV carriers dropped from fifteen percent to one percent after the first ten years of immunization. Although the universal HBV vaccination programs implemented since 1986 have controlled the spread of HBV infection, antiviral drug therapy is required for those chronically infected with the virus to minimize disease progression toward life-threatening sequelae.

3.2.3 Hepatitis B drugs

In addition to hepatitis B vaccines that may be used as prophylactics, several therapeutic drugs are on the market to treat HBV. They may be divided into two classes: (1) interferon that targets the host immune system and (2) six nucleoside drugs that target HBV polymerase. Having the woodchuck as a useful animal model for hepatitis B greatly accelerated the discovery of drugs to treat this infection.

3.2.3.1 *Interferon*
Scottish virologist Alick Isaacs at the National Institute of Medical Research at Mill Hille in north London and his postdoctoral fellow, Jean Lindenmann, discovered interferon in 1957 from their research on virus interference.

Virus interference, discovered in 1935, occurs when, in a cell once infected with one virus (interfering virus), the growth of the second virus is blocked. When an experimental animal was infected simultaneously with two different viruses, sometimes only one of the viruses produced its characteristic disease and infection while the other had no effects. The Henle group in Philadelphia extensively studied virus interference of influenza viruses. They discovered that even an inactivated virus still blocked the multiplication of another. But the mechanism of virus interference was not known until 1957 when interferon was discovered, although as a biological activity (antiviral action), not as a substance. The latter came later.

Isaacs was interested in virus interference for some time. In the summer of 1956 while he was the head of the Chemistry Division at the National Institute of Medical Research, a Swiss postdoctoral fellow, Lindenmann, came to work for him. During their research on virus interference, they made important and consequential observations. If they killed influenza viruses by heating and then added those dead viruses to cells, the cells still resisted subsequent infection with live viruses. Later, they used ultraviolet-inactivated influenza virus to interfere cells from the chorioallantoic membrane of a fertile hen's eggs and obtained even more reproducible results. The dead viruses prompted the cells to resist infection by secreting a substance that Lindenmann called *interferon*, implying its association with *interfer*ence of the virus. Their first publication was not well received by other scientists in the field. Suspecting trace contamination, some virologists derided their results as "misinterpreton" or "imaginon." The accusations deeply saddened Isaacs, who, like many creative people, suffered bouts of severe depression followed by phases of exuberant mental activities. But their subsequent, meticulously executed experiments resoundingly expelled the doubts. Their finding of an innate antiviral response was also the first discovery of a cytokine, long before the interleukins, which allowed the invaded cell to survive.[13]

Isaacs was an extraordinary scientist. He never put his name on a paper unless he had done at least thirty percent of the actual experimental manipulations with his own hands. Sadly, he died in 1967, at age of forty-five, too soon to witness his discovery of interferon gain international recognition. Otherwise, many believed that the significance of the discovery of interferon was on par with many Nobel Prize-winning discoveries.

It was amazing that the interferon was devoid of toxicity, a not unexpected property given that interferon was an endogenous protein. On top of that, interferon had a broad-spectrum antiviral property. It was at a time perceived as a panacea to cure all ills. But its production was a major hurdle. Not only did cells produce interferon in extremely minute quantities, but in most cases the material was species-specific: mice needed mouse interferon and humans needed human interferon. As far as interferon inducers were concerned, Hilleman at Merck discovered that double-stranded RNA could increase the production of interferon exponentially by judicial use of priming and metabolic inhibitors. His best inducer was the complex between polyinosinic and polycytidilic acid (poly I:C), which was very efficient in inducing interferon both in cell culture and in laboratory animals, although not in humans. In the end, the jury is still out on this one. Some touted Hilleman's achievement as ground-breaking, while some others believed that Hilleman dragged the field down the wrong path.

In the early 1960s, Kari Cantell at the State Serum Institute (later renamed the Finnish Blood Center) in Helsinki, Finland set up production of interferon after having spent two years in the Henle group doing interferon research. Although he initially had some success using chick embryo cells, he decided to make interferon from human cells because only human interferon would be suitable to treat patients with various diseases due to species specificity. A breakthrough took place in 1963 when Cantell used white blood cells (leukocytes) obtained from fresh human blood to produce interferon. Using Sendai virus (named after a Japanese city Sendai, 仙台), he was able to boost production exponentially in white blood cells. He also found that serum must be present in the cell culture medium for optimum interferon production. Indeed, in the absence of serum, little interferon was produced by the white blood cells. Cantell and coworkers later discovered that milk could substitute for serum for interferon production. A main protein in milk, casein, could attenuate interferon production just as much as serum did. With production of interferon sorted, Cantell moved onto its purification. In 1972, following the suggestion of Samuel Graff, retired Chair of Biochemistry at Columbia University, Cantell used ethanol to purify interferon and the results succeeded his "wildest expectations." In the following decade, nearly all clinical studies with interferon carried out in all parts of the world used the leukocyte interferon made and purified in Finland by Cantell's group. Indeed, Cantell was deservedly proud of his achievement that raised the international prestige of Finland, a small and remote scientific "backwater." He put Finland on the map of interferon research, the hottest topic in medicine for many years.[14]

Today, most interferon is manufactured using biotechnology, cultured in coliform bacteria. At first, applying the cloning technique, the messenger RNA is extracted and purified from human interferon. The pure messenger RNA can then prepare the corresponding DNA either as a single strand, or in the natural double-stranded form. Gene splicing technique can then insert the genetic information carried by the DNA onto plasmids in the nucleus of a bacterial cell. By harnessing the bacterial cell's own biological production machinery, the cell begins to generate interferon.

Biogen, founded in 1977, was at the brink of bankruptcy in 1980. Luckily, one of the founders, Charles Weissmann, at the University of Zurich, sold his patent on interferon production using the DNA recombinant technology to Schering–Plough for $8 million. Biogen survived and Schering–Plough became the market leader of interferon (trade name Intron). Meanwhile, a team at Genentech led by David Goeddel also succeeded in making recombinant interferon from bacteria. Roche sold the Genentech recombinant interferon with a trade name Wellferon.

Interferon inhibits the multiplication of viruses and regulates cell division, which is why initially it became extremely popular as a potential anticancer drug. Researchers around the world jumped at the opportunity to study it and soon discovered an entire family of these substances, all produced by cells in response to either dead or live viruses. The leukocyte interferon is defined as interferon α, whereas fibroblast interferon is named interferon β. Meanwhile, the "immune" interferon produced mostly from human lymphocytes, discovered by Fred Wheelock in 1965, is now called interferon γ. Interferons α, β, and γ are a family of soluble proteins, belonging to the class of cytokines. Interleukin-2, discovered by Robert Gallo (see Chapter 2), is another one of the cytokines. As a broad-spectrum antiviral drug, it was the only drug to treat hepatitis B at times. Although interferons had a potential anticancer activity, touted by Isaacs himself, it is only efficacious for a particularly rare form of leukemia called hairy cell leukemia.

In 1976, Professor Thomas Merigan at Stanford University published positive results of a small trial of interferon-α in the treatment of four patients with chronic hepatitis B. From then on, more and more clinical evidence accumulated that led to interferon to become the first drug to treat chronic hepatitis B virus infection in 1990 in a recombinant form of interferon-α2 manufactured using gene splicing biotechnology. Nevertheless, the half-time of interferon was too short and the injections had to be given thrice a week. To keep the antiviral interferon staying in the system longer, a polymer was attached to interferon. The polymer was PEG, which stands for *poly*ethylene *g*lycol. Since introduction in 2005, *peg*ylated interferon-α2 has been favored due to the more convenient, once-a-week dosing schedule and a higher response rate resulting from the improved plasma stability. Treatment with interferon leads to undetectable hepatitis B virus DNA levels in a significant percentage of patients with the benefit of lack of emergence of viral resistance. However, the overall cure rate is very low and interferon accounts for no more than ten percent of all prescriptions for the treatment of chronic hepatitis B in the United States. The use of pegylated interferon is further hindered by the extensive appearance of multiple adverse side effects, including influenza-like symptoms, fatigue, anorexia, and emotional instability. Finally, interferon is expensive and administration by subcutaneous injection is not favored by patients either, whereas oral drugs are more convenient and preferred.[15]

3.2.3.2 *Old-fashioned nucleosides*
In addition to interferon, there are six orally administered nucleoside analogs licensed for treating chronic HBV infection. They are (1) the L-nucleosides lamivudine (3TC, Epivir) and (2) telbivudine (LdT), (3) the carbocyclic deoxyguanosine analog

entecavir; (4) adefovir dipivoxil, (5) tenofovir disoproxil fumarate, and (6) tenofovir alafenamide. GSK's lamivudine, Novartis's telbivudine, and BMS's entecavir are simple nucleosides, while Gilead's adefovir dipivoxil, tenofovir disoproxil, and tenofovir alafenamide are acyclic nucleotide phosphonic acid derivative.

Lamivudine (Epivir) was at first approved to treat human immunodeficiency virus (HIV) infection in 1995, and later for treating HBV in 1998. Discovered by Bernard Belleau and coworkers at BioChem Pharma of Canada in 1988, lamivudine was codeveloped with Glaxo Wellcome. It is a reverse transcriptase inhibitor of HIV but has a novel and unique oxathiolane core structure. HBV is a DNA virus and lamivudine exerts its antiviral effects by incorporating lamivudine triphosphate form into viral DNA, which results in DNA chain termination. Clinical trials showed that lamivudine was well tolerated and induced a decrease in serum virus DNA levels associated with normalization of serum *ala*nine amino*t*ransferase (ALT) levels.[16] The ALT levels are routinely screened for blood work during our physical exams where an elevated ALT level is a source of concern regarding liver functions. Long-term consumption of too much alcohol may cause the liver to harden along with elevated ALT levels. The introduction of lamivudine in 1998 as the first oral treatment for chronic hepatitis B ushered in a new era of safe, effective, and well-tolerated drugs. It is used for the treatment of chronic hepatitis B at a lower dose than for the treatment of HIV. It improves the seroconversion of hepatitis B and it also improves histology staging of the liver. Unfortunately, long-term use of lamivudine leads to the emergence of a resistant HBV mutant. Nevertheless, lamivudine is still widely used as it is well tolerated.

Novartis's telbivudine was initially also prepared as an HIV drug, but it was tested to have no activity against HIV. Fortunately, it was found to be quite active against HBV. Novartis gained the FDA approval in 2006 and sold it under the trade name Tyzeka. Telbivudine (Tyzeka) is a *hepatitis B specific* antiviral drug.

An interesting development in the antiviral nucleoside research is the appreciation that unnatural L-nucleosides, mainly cytidine and thymidine analogs, can be phosphorylated by deoxycytidine and thymidine kinases with similar or greater efficiency than their natural D-enantiomers. D-nucleoside are D-ribose sugar derivatives while L-nucleosides are derived from L-ribose sugars. In fact, the L-enantiomer of dideoxycytidine analogs is preferentially favored by deoxycytidine kinase, resulting in more potent antiviral activity compared to the D-enantiomer. Furthermore, L-nucleosides also exhibit more favorable metabolic stability and improved toxicity profiles. As is the case that lamivudine is an L-nucleoside, telbivudine is the unmodified L-enantiomer of the naturally occurring D-thymidine, a pyrimidine deoxynucleoside. Telbivudine and D-thymidine are mirror images of each other.

Telbivudine prevents the virus DNA synthesis by acting as an HBV polymerase inhibitor. Within hepatocytes, telbivudine is phosphorylated by host cell kinase to telbivudine-5'-triphosphate which, once incorporated into the virus DNA, causes DNA chain termination, thus inhibiting HBV replication. In this sense, telbivudine, like most nucleotide antiviral drugs, is a prodrug. Telbivudine is significantly more effective than lamivudine and adefovir. In contrast to other nucleoside analogs, telbivudine has not been associated with the inhibition of mammalian DNA polymerase. Therefore, it is less likely to cause resistance. But resistance eventually materialized nonetheless in the form of HBV signature resistance mutation M204I, which is a change from methionine to isoleucine at position 204 in the reverse transcriptase domain of the hepatitis B polymerase.[17]

The best HBV nucleoside drug is probably BMS's entecavir (Baraclude), which was approved by the FDA in 2005 to treat hepatitis B.

Initially, BMS's antiviral research program targeted nucleosides that would selectively inhibit viral polymerase while minimally interacting with host polymerase. The initial focus was the herpes family viruses, in particular herpes simplex virus (HSV), varicella-zoster virus (VZV), and cyromegalovirus (CMV). At the outset of the project, BMS decided to make dramatic change in their nucleoside analogs' scaffolds. Unlike lamivudine and telbivudine, entecavir has a completely all-carbon carbocyclic core structure. While both lamivudine and telbivudine have a the conformationally flexible 5-membered, deoxyribose ring of the natural nucleosides in a bioactive conformation, in contrast, entecavir has a more rigid cyclopentyl (△) core. The concept seemed outlandish at first because naturally occurring deoxyribose was readily recognized by cellular kinases to phosphorylate it to the corresponding active triphosphate derivative. However, two precedents provided much needed confidence. Nucleoside analogs such as the natural products oxetanocin A and lobucavir were tested positive against HIV-1, HBV, and herpesviruses. Therefore, the mobile tetrahydrofuran moiety of the natural nucleosides could be replaced by a conformationally more rigid four-membered ring that favored a single puckered conformation, represented by the oxetane and cyclobutane (□) moieties in both oxetanocin A and lobucavir. Their being active antiviral nucleosides suggested the feasibility of the approach of using the all-carbon cyclopentyl group as the core structure. Based on structural studies and molecular modeling analyses, the design criteria were used to design lobucavir and ultimately led to the discovery of the hyper-potent carbopentacyclic analog entecavir.

Before becoming entecavir and Baraclude, the compound was known as SQ-34676. It was prepared in the 1990s from a collaboration between the Squibb Institute's Canadian site, Candiac, in Quebec, and its Wallingford site in Connecticut. The compound had been left dormant because it showed little efficacy against herpes simplex virus (HSV). After the merger between Bristol-Myers and the Squibb Institute in 1989, SQ-34676 was still collecting dust in BMS's compound management library although it was given a new name BMS-200475 to reflect the new ownership. On January 13,

1995, the compound's fate changed dramatically when it was tested to have a high potency against HBV—more potent than any previously tested drug. Now the drug had a newly found purpose as a treatment of chronic hepatitis B. Clinical trials demonstrated that BMS-200475 (entecavir, Baraclude) was a clinically effective treatment of HBV infection at an oral dose as low as 0.5 mg per day. In comparison to the other two, old-fashioned nucleoside HBV drugs, entecavir is superior in potency and selectivity, and it has a lower rate of resistance.[18]

3.2.3.3 The Holy Trinity: acyclic nucleotide phosphonates, and prodrugs

In addition to the three previously discussed nucleosides—lamivudine, telbivudine, and entecavir—there are three *acyclic nucleotide phosphonates* HBV drugs. They are adefovir dipivoxil, tenofovir disoproxil, and tenofovir alafenamide, approved in 2002, 2008, and 2016, respectively.

To understand those three acyclic nucleoside phosphonates as HBV drugs, we must look at the genesis of acyclic nucleosides generally first. Acyclic means not a ring. The first acyclic nucleoside drug was acyclovir (Zovirax), discovered by Burroughs Wellcome as a treatment of herpes simplex virus.

In the early 1970s, one of the Wellcome chemists, Howard Schaeffer, established that the intact ribose sugar ring of compounds, such as guanosine (the precursor of acyclovir), was not essential for binding to enzymes needed for DNA synthesis. By cutting off the diol fragment of the sugar fragment, Schaeffer synthesized acyclovir. In contrast to the rigid sugar ring, deannulation of the ribose ring to an acyclic moiety conferred these molecules with the flexibility to present many conformations, one of which was presumably readily recognized by the appropriate host cell kinase to phosphorylate it.[19] Even though Peter Collins and John Bauer, at Burroughs Wellcome in Beckenham, UK, discovered acyclovir's antiviral activities for herpes simplex virus in 1974,

Burroughs Wellcome held back the information for three years. In 1977, Gertrude "Trudy" Elion, who would share the 1988 Nobel Prize with George H. Hitchings and James W. Black, and coworkers reported that acyclovir owed its antiherpetic selectivity to a specific phosphorylation by the herpesvirus-encoded thymidine kinase. A year later, Schaeffer publicly disclosed the antiviral potential of acyclovir, which soon became the "gold standard" for the treatment of herpes simplex virus and varicella zoster virus. Following the discovery of acyclovir, "me-too" drugs including ganciclovir (Cytovene, 1988) and penciclovir (Denavir, 1996) followed. Ganciclovir (Zirgan), discovered by Julien P. Verheyden and John C. Martin at Syntex in 1994, was the first compound with activity against human cytomegalovirus.

Even though acyclovir, ganciclovir, and penciclovir are not HBV drugs, their success as antiviral drugs paved the road to the discovery of adefovir and tenofovir that are HBV drugs.

For a traditional nucleoside antiviral drug to work in the body, the nucleoside must be phosphorylated three times—to become a nucleoside-triphosphate—to be active to insert into the DNA chain to serve as a DNA chain terminator. The first step to phosphorylate the nucleoside with a kinase is very slow. Therefore, it was conjured that installing a phosphate to the nucleoside would accelerate the "activation" process. A nucleoside with one or more phosphates attached is known as a nucleotide. In theory, a nucleotide is more advantageous than a nucleoside drug. The problem is that the phosphate group on nucleotide is readily hydrolyzed by phosphatases because the P–O bond is very weak. In contrast, the P–C bond is exponentially more robust and non-biodegradable. Consequently, a phosphonate is significantly more resistant to hydrolysis by phosphatase than the corresponding phosphates.

In 1986, a collaboration between Antonín Holý of Academy of Sciences in Prague, Czech Republic and Erik De Clercq of Rega Institute at Leuven, Belgium disclosed for the first time the merit of acyclic nucleotide phosphonates. The phosphonate derivatives adefovir and tenofovir are representatives of this tactical approach to drug design. The phosphonate acts as a mimic of a nucleoside monophosphate that effectively circumvents the first phosphorylation step (known as *kinase bypass*) and confers stability toward hydrolysis by phosphomonoesterases. The molecular flexibility associated with these two compounds has also been postulated to be a key factor contributing to the suppression of the development of virologic resistance during antiviral treatment.

The polar phosphonate group of acyclic nucleotide analogs is frequently associated with kidney toxicity resulting from the accumulation of the drug in the renal proximal tubules, a toxicity profile that limits the dose of adefovir that can be used in the clinic for treating chronically HBV-infected patients. Too much polarity of the phosphonate group found in adefovir and tenofovir also prevents efficient absorption across the lipid bilayer of the intestine that is essential for oral exposure. This is remedied by masking the phosphonate group with the prodrug elements. The prodrug strategy can sometimes turn a terrible drug into a decent one— transforming an ugly duckling to a beautiful swan. The litany of a prodrugs's merits includes:

- Overcoming formulation and administration problems,
- Overcoming absorption barriers,
- Overcoming distribution problems,
- Overcoming metabolism and excretion problems, and
- Overcoming toxicity problems.

Prodrugs currently constitute five percent of known drugs and a larger percentage of new drugs. Again, the prodrug strategy is not a panacea.

Adefovir dipivoxil and tenofovir disoproxil are prodrugs also discovered by Holý and De Clercq. Initially, Holý lobbied to have Syntex to license their phosphonate drugs, but Syntex stuck with their own ganciclovir (Zirgan). De Clercq enticed Bristol-Myers for development of the drug. He visited Bristol-Myers's Wallingford site in Connecticut often and befriended chemist John C. Martin. The merger between Bristol-Myers and the Squibb Institute in 1989 resulted in an exodus of a cadre of managers, Martin included, who joined Gilead Sciences in Foster City, California. For portfolio consolidation reasons, the newly formed Bristol-Myers Squibb chose to develop oxetanocin and entecavir and sold the intellectually properties on adefovir dipivoxil and tenofovir disoproxil to Gilead at a steeply discounted price. Gilead then shepherded the two drugs through clinical trials. Adefovir dipivoxil was initially developed as a treatment for HIV, but the FDA in 1999 rejected the drug due to concerns about the severity and frequency of kidney toxicity when dosed at high doses. Nevertheless, adefovir dipivoxil was effective at a much lower dose for the treatment of chronic hepatitis B in adults with evidence of active viral replication and either evidence of persistent elevations in serum alanine aminotransferases (ALT) or histologically active disease. Overall, the efficacy of adefovir dipivoxil against wild-type and lamivudine-resistant HBV and the delayed emergence of adefovir dipivoxil-resistance during monotherapy contribute to the durable safety and efficacy observed in a wide range of chronic hepatitis B patients.

Tenofovir is a nucleotide analog closely related to adefovir with just an additional chiral methyl group. *In vitro* studies showed that it had activity against HBV with equimolar potency to adefovir. Clinical studies confirmed its efficacy in suppressing HBV replication, and it appears to be equally effective against both wild-type and lamivudine-resistant HBV.

The two prodrugs are stable during the passage through the intestine and easier to penetrate the cell membrane to get into the cells. Once in the cells, especially liver cells, esterases can then cleave the prodrug fragments and reveal the active, polar acyclic nucleoside phosphonates that can exert their pharmacological activities by serving as DNA chain terminators. The prodrug adefovir dipivoxil has a thirty percent improvement of bioavailability, which

translates to an approximate three-fold improvement in oral bioavailability in humans. Adefovir dipivoxil (Hepsera) was approved
to treat chronic hepatitis B in 2002. Tenofovir disoproxil (Viread)
was approved as a treatment of HIV in 2001 and as a treatment of
chronic hepatitis B in 2008. It became the gold standard for the
treatment of both HIV and HBV.[20]

This class of acyclic nucleoside phosphonate prodrugs are called
the *Holý Trinity* by the United Nations Educational, Scientific and
Cultural Organization (UNESCO). The three scientists largely responsible for their discovery and development are Holý, Martin,
and De Clercq (see Figure 3.5).[21]

In 2016, Gilead gained approval of tenofovir alafenamide
(Vemlidy), the only approved HBV drug during the last decade.
Considering none of the five previous nucleoside HBV drugs
represented a cure, a new drug must show clear differentiation in
terms of improved efficacy, high genetic barrier to mutation, and
a better safety profile. Tenofovir alafenamide surpasses those really high hurdles. The trick is switching the prodrug portion of
tenofovir disoproxil to a superior phosphoramidate prodrug

Fig. 3.5 Antonín Holý
© Czech Post

known as ProTide, which stands for *Pro*drugs of nucleo*Tides*. Having taken a page from their success of the HCV drug sovaldi (*vide infra*), Gilead installed the same ProTide prodrug onto tenofovir to make tenofovir alafenamide, which demonstrated similar efficacy to tenofovir disoproxil at a mere twenty-five mg daily dosage, one-tenth of the latter. It has improved kidney and bone safety due to increase exposure of active form of tenofovir in liver and reduced systemic circulation of tenofovir.[22]

All six previously discussed nucleoside analogs have been shown to suppress viral replication, as measured by serum HBV DNA levels. These drugs resulted in enhanced treatment outcomes, including the normalization of serum alanine aminotransferase (ALT) levels and improvements in liver histology. Although virologic resistance remains a concern with nucleoside analog therapy, newer agents approved subsequent to lamivudine, particularly entecavir, adefovir dipivoxil, tenofovir disoproxil, and tenofovir alafenamide have shown much improved resistance profiles.

In summary, the six nucleoside HBV drugs on the market are:

- lamivudine (3TC, Epivir, GSK) in 1998,
- adefovir dipivoxil (Hepsera, Gilead) in 2002,
- entecavir (Baraclude, BMS) in 2005,
- telbivudine (LdT, Tyzeka, Novartis) in 2006,
- tenofovir disoproxil (Viread, Gilead) in 2008, and
- tenofovir alafenamide (Vemlidy, Gilead) in 2016.

3.3 Hepatitis C Virus

Hepatitis C is caused by the blood-borne hepatitis C virus (HCV). The WHO estimates that there were seventy-one million people chronic hepatitis C infection worldwide in 2017 although there were 170 million back in 1999, about three percent of the world's population. Each year, 350,000–500,000 people die from liver cirrhosis, liver cancer, and hepatocellular carcinoma, which then require liver transplantation. HCV causes at least eighty-five percent of the cases of transfusion-associated hepatitis. Preventative measures, including use of disposable syringes and screening of blood used for transfusions, have reduced the incidence of new HCV infections in recent years, but disease transmission still occurs in healthcare settings that do not practice adequate disease control measures. The virus is also spread in non-medical settings through the use of contaminated intravenous drug, piercing, and tattooing equipment, and through unprotected sex. Because the infection is often "clinically silent" for years, most infected people are unaware of their infection.

Unlike AIDS and hepatitis B, hepatitis C is a curable disease. Its cure rates have improved dramatically over the past twenty years. "Cure" is defined as eradication of virus and maintenance of a sustained viral response (SVR), without detectable virus RNA, for six months after completion of treatment.

3.3.1 Discovery of hepatitis C virus

HCV was discovered in 1989, merely some thirty years ago from today. Opportunely, when the Nobel Prize was awarded in 2020, many cures already existed for hepatitis C. On October 5, 2020, the Nobel Assembly at the Karolinska Institute announced that the Nobel Prize for Physiology or Medicine was awarded to Harvey J. Alter, Michael Houghton, and Charles M. Rice "for their discovery of hepatitis C virus." The discovery of HCV paved the road for blood testing and finding effective treatments, which now represent a cure. This was a resoundingly worthy award. All three awardees that year had previously received the Lasker Awards, an American award widely viewed as a forerunner of the Nobel Prize. When Alter was woken up by phone calls from Stockholm at midnight, he was annoyed and did not pick up until the third time. Apparently, he was not waiting by the phone for the call.

The award to HCV discovery is in line with the Nobel Committee's long tradition of awarding the discovery of a major virus. To be sure, the Nobel Foundation rarely gives out prizes for discovering a virus until a treatment, a cure, or a vaccine has been found. Otherwise, awarding the discovery of a killer virus would be viewed as "bad form." No doubt the fact that the ongoing COVID-19 epidemic caused by SARS-CoV-2 was wreaking havoc around the world provided additional incentive for Stockholm to favor an award for virology.

Here is the list of all virus-related Nobel Prizes:

- 1966, Discovery of Rous sarcoma virus, Peyton Rous
- 1976, Discovery of HBV, Baruch Blumberg
- 2008, Discovery of HIV, Luc Montagnier and Françoise Barré–Sinoussi
- 2008, Discovery of HPV, Harald zur Hausen
- 2020, Discovery of HCV, Harvey Alter, Michael Houghton, and Charles Rice

Back in 1975, Alter and coworkers at NIH found that most cases of transfusion-associated hepatitis did not have the serological marker of either HAV or HBV. He called the "new" disease "non-A, non-B viral hepatitis." A frustrating stretch of fourteen years began, during which the methods used to successfully identify HAV and HBV all failed to reveal the molecular identification of the etiological agent of this disease. The traditional techniques that they used to identify the virus included tissue culture growth, electron microscope, and serological identification. The causative agent, later named as hepatitis C virus (HCV), was eventually discovered in 1989 by Michael Houghton and coworkers using exclusively a molecular biology technique. To which Alter commented that, "the cart has led the horse": *Descartes before the horse!*[23]

Houghton's group at Chiron Corporation accomplished the cloning of a large portion of the HCV genome with a large volume of chimpanzee plasma provided by Dan Bradley at the Centers for Disease Control and Prevention (CDC). The chimpanzee plasma had shown to have an unusually high infectivity titer. The cloning of HCV was initiated by ultracentrifugation of the chimpanzee plasma. Genomic characterizations through cloning revealed that the virus was an RNA virus related to the *flavivrus* family. This discovery represented, for the first time, that a new virus was identified using exclusively molecular biology techniques from the blood of an experimentally infected chimpanzee. Immediately following the cloning of the virus, Houghton and a colleague, Qui-Lim Choo, developed a diagnostic assay for detecting HCV antibodies. Second cloning in yeast resulted in an antigen from three contiguous genomic segments (epitopes). This led to the development of the first-generation enzyme-linked immunosorbent assay (ELISA). Later, more sensitive enzyme immunoassays and molecular biology-based assays were developed to detect HCV RNA in body fluid. In Houghton's lab, Choo provided many years of outstandingly dedicated and precise molecular biology expertise. Meanwhile, Houghton's colleague George Kuo provided intellectual and practical input. Houghton treasured his coworkers' contributions to the discovery of HCV. In 2013, when he was bestowed a $100,000 award by the Gairdner Foundation, he turned down Canada's most prestigious science award because his close collaborators, Choo and Kuo, were not included. Currently, Houghton is a professor at the University of Alberta. His Nobel Prize ended the century-long draught of Canadian's Nobel Prize in Physiology or Medicine. The last Canadian who won the accolade was Frederick Banting in 1923 for his discovery of insulin.

Confirmation of the accuracy and fidelity of the Chiron's diagnostic assay was carried out in Alter's group at the NIH. The Chiron test detected all but one of the previous putative HCV sera in the panel and had no discordant results on duplicate analysis and no false positive results. The one sample missed was obtained during the acute phase of the disease and, thus, during the window period before antibodies developed.[24]

Unlike "naked" HAV, HCV has a lipid envelope. But like HAV, HCV is a single-stranded positive-sense RNA virus. Like RNA viruses in general, HCV does not replicate through a DNA intermediate; therefore, it can be eradicated. There are six major HCV genotypes.

HCV therapy initially relied on treatment with the immune stimulant interferon. Subsequently improved by the addition of daily oral doses of the nucleoside analog ribavirin. Ribavirin was discovered in 1972 by Robins and colleagues at the International Chemical and Nuclear (ICN) Corporation in Irvine, California. Structurally, ribavirin is a nucleoside analog in which the natural base is replaced by a carboxamidotriazole fragment. It was found to be a broad-spectrum antiviral drug, active against at least sixteen DNA and RNA viruses. Ribavirin (Virazole) was approved by the FDA in 1986 as a treatment of respiratory syncytial virus (RSV). In 1991, Swedish scientists at the Karoliska Institute reported a pilot study with oral ribavirin as therapy for chronic hepatitis C. Even though their data was not robust, ribavirin's potential as an HCV drug generated a lot of interest. The combination of ribavirin with interferon was approved in 1998 by the FDA for the treatment of HCV, which became the standard of care (SOC) for treating hepatitis C before the emergence of more selective direct-acting anti-HCV drugs.[25] Later on, the introduction of pegylated interferon-α reduced the dosing frequency of the protein to once-weekly subcutaneous injections. This combination improved response rates. As mentioned before, the most significant side effects associated with pegylated interferon-α are flu-like symptoms. Additionally, ribavirin may cause hemolytic anemia. In many cases, the severity of these side effects caused patients to discontinue the treatment.

3.3.2 Hepatitis C virus NS3/4A serine protease inhibitors

Charles Rice, at Washington University at the time, completed the characterization of the HCV viral genome in 1996 and a year later succeeded in producing an infectious virus in the lab. He went on to develop sub-genomic amplicons of the virus that could replicate in cells without producing live virus, which made it possible to design assays to test for drugs capable of directly inhibiting viral replication.[26]

The isolation and detailed characterization of the HCV in 1989 set the stage for a drug discovery campaign focused on discovering and developing direct-acting antiviral agents that specifically target viral proteins and nucleic acid. These HCV inhibitors were anticipated to offer considerable potential for improved response rates with therapy more tolerable than the combination of interferon and ribavirin. There was an ulterior motive for big pharma to work on HCV drugs. While HBV is largely a disease of developing countries, HCV inflicts many developed countries, which can afford paying for expensive innovative drugs.

Since the discovery of HCV, academia, government agencies, and especially the pharmaceutical industry have invested a tremendous number of resources and have discovered many effective anti-HCV drugs. A cure can now be achieved as measured by sustained virologic response (SVR). Eradication of HCV may be achieved in the foreseeable future.

Rice's discoveries were instrumental to the discovery of anti-HCV drugs. Figure 3.6 illustrates a portion of the HCV genome associated with non-structural (NS) proteins. Nearly all HCV non-structural proteins have proven to be valid targets for inhibiting HCV replication.

The HCV NS3 serine protease is one of several enzymes required for HCV replication in humans. The NS3 protein has two components: an N-terminal serine protease and a C-terminal

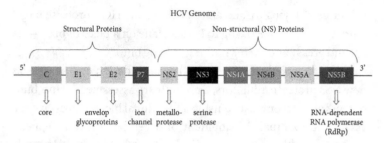

Fig. 3.6 The HCV Genome

RNA helicase. It was found to be a trypsin-like serine protease with a molecular weight of seventy-five kDa. Functionally, the HCV NS3 serine protease is responsible for cleaving the viral polyprotein at four sites including the NS3-4A, NS4A-4B, NS4B-5A, and NS5B-5B junctions. In other words, the HCV NS3 protease enzyme cuts the large *non-functional* polypeptides into smaller *functional* proteins. Inhibiting this enzyme would stop the virus' replicating process. The stability and activity of NS3 protease is enhanced by a short peptide cofactor, NS4A. For NS3 protease to be active as a protease, it must be complexed with NS4A. This is why NS3 inhibitors are mutually interchangeable with NS3/4A inhibitors.

The groundbreaking discoveries of successful HIV protease inhibitor drugs suggested that the HCV NS3 serine protease might be a viable target for design of new, oral anti-HCV drugs. The enzyme's solid-state structure was determined using X-ray crystallography, which helped to launch structure-based drug design (SBDD) efforts for this target. However, this crystal structure also revealed that, unlike the deep, "druggable" substrate-binding pocket of HIV protease, the substrate-binding region of HCV NS3 is a shallow, solvent-exposed groove. From the aspect of drug design, NS3 thus presented a significant challenge to medicinal chemists in the design of selective and potent small-molecule inhibitors of this target. A Vertex chemist has captured the difficulties posed by this target: "Trying to land an inhibitor in the HCV protease target binding site was like trying to land a plane on a piece of pizza—it's flat and greasy and there's nothing to hang onto."[27]

As medicinal chemistry teams began to design and synthesize new NS3 protease inhibitors, a preliminary assessment of inhibitor potency was accomplished using a functional biochemical enzyme assay. The eventual development of a cell-based HCV replicon assay enabled the assessment of new inhibitors in a physiologically cellular environment, thus providing a major advance in discovery efforts for this target.

Boehringer Ingelheim was the first company to put an HCV NS3 protease inhibitor to clinical trials in humans. Discovered at their Quebec site in Canada, Boehringer Ingelheim used the protease cleavage *products* as their starting point. Their groundbreaking drug candidate ciluprevir, a macrocycle, established clinical proof of concept for this class of direct-acting antiviral drugs. In clinical trials for ciluprevir, following two days of dosing to HCV-infected subjects, more than a hundred-fold decline of HCV RNA in their blood was observed. When they stopped taking ciluprevir, their HCV RNA level returned. Regrettably, Phase Ib clinical trials for ciluprevir were terminated around 2004 due to cardiotoxicity issues observed in Rhesus monkeys at high doses. Apparently, the cardiotoxicity was molecule-specific for ciluprevir, not mechanism-based because many compounds of the same chemotype later are found to be devoid of such cardiotoxicity. Boehringer Ingelheim's ciluprevir remains a seminal contribution to the field of NS3/4A protease inhibitors. Its follow-up drug candidate, faldaprevir, a tripeptide-based NS3 inhibitor, was devoid of the cardiotoxicities that plagued the prototype ciluprevir. It was an acyclic molecule that removed the challenge of conducting ring-closing metathesis-mediated macrocycle construction at scale. Regrettably, Phase III clinical trials for faldaprevir were terminated in 2014 for business reasons because many HCV treatments from competitors had already become available on the market at the time. Despite being a pioneer of the field, Boehringer Ingelheim has no HCV NS3 inhibitors on the market.[28]

3.3.2.1 *Slowly reversible covalent inhibitors*

The NS3 protease inhibitors boceprevir and telaprevir were the first HCV direct-acting antiviral agents to be approved in May 2011 for treating HCV patients in combination with pegylated interferon and ribavirin. They are both tetrapeptide derivatives that rely upon extensive contacts between the inhibitor and protease in addition to the formation of a reversible covalent bond between the catalytic serine-139 and the activated carbonyl moiety that is embedded in the α-keto amide element. These agents are slow-binding, mechanism-based inhibitors of the enzyme that form stable but reversible complexes with the protein.

To address the challenging substrate-binding site of HCV NS3/4A, a number of research teams began to explore the design of reversible covalent inhibitors. This strategy, which produced drug candidates for other protease inhibitors, involved the incorporation of an electrophilic trap, or "warhead," into a substrate-like inhibitor. Reversible nucleophilic addition of the NS3 protease catalytic-site serine upon an electrophilic warhead group could confer considerable potency and selectivity advantages to the inhibitor, enabling the design of low molecular weight drug candidates with attractive pharmacokinetic properties.

Schering–Plough was one of the Big Pharma before being gobbled up by neighboring Merck in 2009. It already marketed the first pegylated recombinant interferon (Intron A) as the first approved treatment of HCV at the turn of the twentieth century. Therefore, hepatitis C was a disease of great interest to the company. They chose to find HCV NS3 serine protease inhibitors because of the crucial role that the enzyme played in the lifecycle of the virus. In addition, Schering–Plough, an industry leader of structural biology, had had a lot of experiences in tackling serine proteases. They felt that they were well positioned to be successful. From their high throughput screening campaign, they were somewhat surprised and very much disappointed: screening their corporate compound library of four million compounds did not identify *any* viable lead. They came out of it completely empty-handed.

Without any lead compound as the starting point, Schering–Plough decided to pursue a structure-based drug design approach. In contrast to Boehringer Ingelheim, Schering–Plough used the protease *substrate* in the form of a undecapeptide, a peptide with eleven amino acids, as their starting point. For protease inhibitors, aldehydes, trifluoromethyl ketones, boronic acids, fluoroketones, ketoheterocycles, ketoacids, and ketoamides were well-known "warheads" that served as reversable electrophilic traps of the enzyme. After aldehyde and trifluoromethyl ketone warheads failed to work out, Schering–Plough chose ketoamides as warheads to incorporate onto the natural substrate of the enzyme. To the undecapeptide was installed a ketoamide warhead as the electrophilic trap, which would form a covalent bond between the catalytic serine-139 and the activated carbonyl moiety. The ketoamide warhead would become the hallmark of both boceprevir and telaprevir. An undecapeptide was not suitable to become a drug because a peptide could not be readily absorbed but would be rapidly metabolically hydrolyzed to individual amino acid building blocks and lose its activity. Meticulous and methodical engineering to reduce the molecular weight and modifying each amino acid led to the discovery of boceprevir that was potent and orally bioavailable.[29]

After boceprevir (Victrelis) was approved by the FDA in 2011, Schering–Plough/Merck kept optimizing their series of ketoamide NS3/4A inhibitors. One of them, narlaprevir, was ten-times more potent than boceprevir. It was approved in Russia in 2016 and Merck sold it under the trade name Arlansa.

Vertex's NS3 inhibitor, telaprevir (Incivek) won the FDA's nod only ten days after Schering–Plough's boceprevir (Victrelis) in May of 2011. Boceprevir has a similar ketoamide warhead serving as an electrophilic trap.

As early as 1993, Vertex became interested in tackling the HCV by establishing a liaison with Charles Rice, who was then just an assistant professor at Washington University. One of Vertex's justifications to work on HCV was that the company was already familiar with proteases from their experiences with the HIV protease. Their exploits with HIV protease amprenavir (Agenerase) were covered in Section 2.3.1. In the summer of 1996, Vertex successfully grew their first HCV NS3 protease diffraction-grade crystals. The enzyme had a very smooth surface, lacking the usual cleft that caused binding sites to be buried in a pocket. A virologist from Schering–Plough, Ann Kwong, joined Vertex in 1997 and brought with her a lot of experiences in HCV biology. Meanwhile, Vertex began an alliance with Eli Lilly to work on HCV NS3 protease with Lilly footing the $50 million price tag. Lilly screened their entire compound management library against the NS3 protease and came out completely empty-handed with no verified hit, just like Schering–Plough did. This left them with no choice but to pursue the structure-based drug design approach, also just like Schering–Plough did. Vertex chemists led by Roger Tung designed new scaffolds and warheads. They used a hexapeptide scaffold disclosed by Boehringer Ingelheim as their starting point and they initially installed an aldehyde warhead also without success. But the ketoamide warheads, the same as that of Schering–Plough's, proved to be ideal. The collaboration eventually led to the discovery of telaprevir.[30]

Lilly process chemists were appalled at the drug's physiochemical properties. It was too big, too greasy, and too insoluble. Highly crystalline, telaprevir had as good a solubility as that of brick dust. In addition, it cost $2.5 million to make just one kilogram of the drug, requiring twenty-two steps with very low yields for many of the steps. Lilly decided to discontinue their collaboration with Vertex on this project. A vote of no-confidence by Lilly did not waiver Vertex's confidence on the compound. One source of their confidence was that the drug preferentially distributed to the liver, where it was supposed to be. Clinical trials began in 2004 although the company was still struggling to formulate the drug to an amorphous form with polymers to enhance its solubility during the Phase Ib trials. During the Phase II trials, Vertex signed a co-development deal with Tibotec, a division of Johnson & Johnson, with the pharma giant footing all development bills and nearly 500 million dollars of combined milestone payments to the smaller biotech partner.[31]

The approval of boceprevir and telaprevir in 2011 as the first direct-acting antiviral drugs to treat hepatitis C marked the start of a new era in treatment paradigms. Vertex's telaprevir became the fastest drug to reach $1 billion in cumulative sales in just over six months. Interesting, many better HCV drugs quickly followed on the market. By the end of 2014, Vertex announced that it would discontinue the sale and distribution of telaprevir (Incivek) in the United States due to financial considerations. The same fate also befell Schering–Plough's boceprevir (Victrelis) around the same time.

3.3.2.2 Macrocyclic reversible NS3/4A inhibitors

Boehringer Ingelheim's ciluprevir and faldaprevir built the foundation for the success of all ensuing NS3/4A inhibitors. Moreover, the chemotype for ciluprevir and faldaprevir became the prototype of "me-too" drugs that took advantage of their pharmacophore to make successful drugs by simple modifications to minimize the toxicity and boost the potency.

Using Boehringer Ingelheim's macrocyclic NS3/4A inhibitor faldaprevir, Medivir/Janssen were the first to gain the FDA approval of a macrocyclic reversible NS3/4A inhibitor in 2013 with simeprevir. Several features of simeprevir differ from faldaprevir. At the outset of the project, Medivir in Huddinge, Sweden decided to use cyclopentane (△) core structure to replace the prototype ciluprevir's proline core, no doubt motivated by the desire to have a metabolically more robust scaffold. A Janssen subsidiary, Tibotec in

Ireland, in collaboration with Medivir, employed acylsulfonamide as the bioisostere of the carboxylic acid. Their identification of the *cyclopropyl (Δ) acylsulfonamide* moiety as a structural motif that preserved the acidic element and conferred potency enhancements of up to one-thousand-fold compared to simple cyclopropyl carboxylic acid analogs present on both ciluprevir and faldaprevir. It is not surprising that all following "me-too" NS3/4A inhibitors contain this magic cyclopropyl acylsulfonamide motif. Eventually, Medivir/Tibotec/Janssen's simeprevir (Olysio) was approved in 2013. In contrast to the two previously approved NS3/4A inhibitors telaprevir (Incivek) and boceprevir (Victrelis) that are slowly reversible covalent inhibitors, simeprevir (Olysio) is a garden variety reversible NS3/4A serine protease inhibitor, and so are all follow-up "me-too" drugs.[32]

A nomenclature carton is shown in Figure 3.7 to better orient us regarding protease and protease inhibitors. In essence, a protease normally has a catalytic metal ion, Zn^{++}, for instance, at the enzyme's active site. Arbitrarily, we define the first binding pocket on the left of the active metal ion as S_1 and the second on the left as S_2, so on and so forth. The first binding pocket on the right of the catalytic metal is defined as S_1' and the second one on the right as S_2'. For the endogenous ligand in the form as a peptide chain, the fragment occupying the S_1 pocket is defined as P_1 region. Meanwhile,

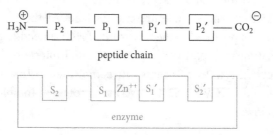

Fig. 3.7 Nomenclature of Porteases

the peptide fragment that occupying the S$_1$' pocket is known as the P$_1$' region.

One case in point is Boehringer Ingelheim's ciluprevir. Its macrocycle is formed by the P1–P3 region, as are Janssen's simeprevir and Roche's danoprevir (Ganovo, Roche, 2013, only in China). On the other hand, all other follow-up reversible HCV NS3/4A serine protease inhibitors are P2–P4 macrocycles. They include Merck's vaniprevir (Vanihep, 2014, only in Japan), grazoprevir (Zepatier, combination with NS5A inhibitor elbasvir, 2016), and narlaprevir (Arlansa, 2016, only Russia); Abbvie's glecaprevir (Mavyret, combination with NS5A inhibitor pibrentasvir, 2017) and Gilead's voxilaprevir (Vosevi, combination with sofosbuvir and ledipasvir, 2017). For instance, voxilaprevir has favorable binding interactions with the HCV NS3/4A protease's Ser-His-Asp (S139, H57, and D81) catalytic triad that is conserved across all genotypes and R155, and thus has a picomolar potency for genotype 3 and genotype 1 resistance-associated substitutions. Vosevi provides a treatment option for the most difficult-to-cure patients who have failed prior direct-acting antiviral-based therapy. For direct-acting antiviral-naïve patients, Vosevi offers a pan-genotypic option of shortened treatment duration of eight weeks. Considering HCV has eight genotypes and at least eighty-six viral subtypes, pan-genotypic drugs are ideal for a cure.[33]

Closely related to Boehringer Ingelheim's faldaprevir, BMS's asunaprevir is an *acyclic* reversible HCV NS3/4A serine protease inhibitor on the market. Regrettably, asunaprevir (Sunvepra) has hepatoxicity warnings on its prescription label. All these drugs offer considerable potential in the treatment of HCV infection as part of combination therapies with either pegylated interferon/ribavirin or direct-acting antiviral agents with complementary mechanisms.

Marketed HCV NS3/4A serine protease inhibitors are listed below:

- boceprevir (Victrelis, Schering-Plough/Merck) in 2011,

- telaprevir (Incivek, Vertex) in 2011,
- simeprevir (Olysio, Tibotec/Medivir/Janssen) in 2013,
- danoprevir (Ganovo, Roche) in 2013 in China,
- asunaprevir (Sunvepra, BMS) in 2014,
- vaniprevir (Vanihep, Merck) in 2014 in Japan,
- grazoprevir (a component of Zepatier, in combination with NS5A inhibitor elbasvir, Merck) in 2016,
- narlaprevir (Arlansa, Schering-Plough/Merck) in 2016 in Russia,
- glecaprevir (a component of Mavyret, in combination with NS5A inhibitor pibrentasvir, Abbvie) in 2017, and
- voxilaprevir (a component of Vosevi, in combination with sofosbuvir and velpatasvir, Gilead) in 2017.

3.3.3 Hepatitis C virus NS5A protein inhibitors

Unlike many proteins in the HCV genome, such as NS3 protease and NS5B polymerase, the nonstructural protein 5A (NS5A) actually has no enzymatic activity. But that does not mean NS5A is useless. In fact, it is required for HCV RNA replication and virion assembly and NS5A is called the master regulator of the HCV life cycle. The crystal structure of NS5A protein was published in 2005. Despite the difficulty in finding a drug not associated with enzymatic activity, BMS's daclatasvir was the first NS5A inhibitor to show efficacy in clinics in 2008. Single doses of the drug effected pronounced and rapid decline in viral RNA in HCV-infected patients.

In a classic application of forward chemical genetics, BMS used a *phenotypic* replicon assay to screen over a million compounds and identify an iminothiazolidinone as a modestly potent inhibitor of HCV. Experimentation and analysis indicated that the hit worked by binding to the NS5A protein. Even though the potency of the hit from high-throughput screening (HTS) could be easily improved by more than 250-fold by simple removal of an oxygen, the chemotype was not stable in some organic solvent, degrading via a radical pathway. A critical and enlightening experiment discovered that a dimer, formed from the radical degrading process in the assay medium, was a potent inhibitor in a replicon assay. Subsequently, the dimer was simplified as a symmetrical molecule with a stilbene core structure, which then became a new lead compound for a group of chemists at the BMS Wallingford site led by Nick Meanwell. Daclatasvir emerged from a significant iterative campaign of structural refinement focused on enhancing potency, expanding the virology profile to encompass a broad range of HCV genotypes, and optimizing pharmacokinetic properties to afford a compound that was active in cell-based assays at picomolar concentrations. Their efforts culminated in the discovery of a hepatitis C virus NS5A replication complex inhibitor daclatasvir (Daklinza). Approved in 2015, it became the first drug to validate translation of phenotypic replicon to a clinical effect.[34] BMS's daclatasvir had an indelible impact on other NS5A inhibitors. In fact, structural features of daclatasvir can be found each of all the four following "me-too" NS5A replication complex inhibitors including: Gilead's ledipasvir and velpatasvir; Merck's elbasvir, and Abbvie's ombitasvir and pibrentasvir.

Interestingly, Gilead's ledipasvir was approved by the FDA in 2014, one year ahead of BMS's daclatasvir (Daklinza) despite Gilead's late start. Impressively, when Gilead started working on HCV NS5A inhibitor inhibitors, the company had over twenty HCV research programs ongoing and several compounds undergoing, or selected to enter, clinical trials. The structure of ledipasvir bore some resemblance to BMS's daclatasvir but with some differentiations. It was very likely that Gilead gained structural insight from BMS's patents on daclatasvir. There was nothing untoward about this practice, which is often colloquially called *patent busting*. One of the major purposes of patents, in addition to preventing competitors from doing exactly the same thing, is to encourage competition, as long as there is enough differentiation. For instance, daclatasvir has a C_2 symmetry, whereas ledipasvir is unsymmetric. Gilead found that their unsymmetric approach afforded intriguing structural variation and superior properties. Although both drugs have a bi-phenyl moiety, ledipasvir's core structure is a fused tricyclic difluorofluorene. The two fluorine blocked the metabolic soft spot on the fluorene ring, a tactic routinely deployed these days. In the end, it was also found that a spirocyclopropyl modification of the pyrrolidine fragment on the left of the molecule provided the most potent inhibitor. The resulting drug ledipasvir has a picomolar antiviral potency. Considering it is ninety-nine percent protein bound, the actually potency is as high as femtomolar! Ledipasvir also has a long pharmacokinetic half-life of over forty hours. Gilead cleverly added their blockbuster drug sofosbuvir to ledipasvir as a combination drug with a trade name Harvoni. The drug was the first approved single-tablet regimen for the treatment of HCV infection with a stunning cure rate of ninety-five percent, with a treatment duration as short as eight weeks.[35]

After completing the discovery of ledipasvir, Gilead moved on to discover an NS5A inhibitor with the potential to effectively treat all HCV-infected patients, regardless of genotype. The culmination of their efforts led to the discovery of the pan-genotypic NS5A inhibitor velpatasvir with high potency *across all six genotypes*. It was combined with their NS5B inhibitor sofosbuvir to become a single-tablet regimen Epclusa, which was approved in 2016. Velpatasvir was also combined with Gilead's superstar NS5B inhibitor sofosbuvir and NS3/4A inhibitor voxilaprevir to make another single-tablet regimen Vosevi, which was approved in 2017.[36]

Merck pursued HCV targets for three decades. Its NS5A inhibitor elbasvir was discovered by their newly formed External Basic Research (EBR) group and a team of scientists from WuXi AppTec. In 2008, the EBR-8 group, led by Craig Coburn, started the HCV NS5A project with the support of twenty-five chemists from WuXi AppTec in China. Even though Merck's initial piperazine-based hits were associated with NS5A, the structures begun to slowly morph into daclatasvir-like molecules incorporating imidazoles, pyrrolidines, and L-valines, no doubt strongly influenced by BMS's success. Merck was able to obtain novel intellectual properties when they used an indole ring in place of daclatasvir's phenyl ring. Moreover, the unique tetracyclic morpholine-like scaffold was also innovative. The fruit of their labor led to the discovery of elbasvir. It was combined with Merck's MS3/4A protease inhibitor grazoprevir and the single-tablet regimen was approved in 2016 with a trade name Zepatier.[37]

A giant in medical diagnostics, Abbott was one of the first companies to commercialize a blood screen test for HCV. Abbvie, spun off from Abbott in 2013, has had a long tradition in the field of infectious diseases, having made great contributions in antibacterial and antiviral drugs. Abbvie's NS5A inhibitor pibrentasvir owed its genesis to an HTS hit from a *phenotypic screen*. While being agnostic to mechanism of action, the hit compounds inhibited replication of the HCV genotype 1b subgenomic replicon. One of them, a naphthyridine emerged as a potent inhibitor genotype 1b replicon. Abbvie had great success in discovering HIV protease inhibitor ritonavir (Norvir) using the C2-symmetry strategy, which also worked well here for their NS5A inhibitor program. Their extensive medicinal chemistry efforts led to the discovery of ombitasvir. Its efficacy and safety data from Phase III clinical trials justified regulatory filing. Abbvie combined ombitasvir with their NS3/4A protease inhibitor paritaprevir, NS5B polymerase inhibitor, dasabuvir and pharmaco-enhancer ritonavir and has sold the combination of four drugs with a trade name Viekira Pak since 2014. The combination drug worked very well but having *four drugs* in one pill is getting close to a witch's brew, making its pharmacology and drug–drug interactions more complicated. It would be vastly better if fewer drugs could accomplish the same goal. To further improve upon the first-generation NS5A inhibitor ombitasvir, Abbvie made three major modifications to the molecule. The isopropyl substitutes were replaced by two ether analogs, which presumably reduced the lipophilicity of the drug. Another greasy t-butyl-phenyl "cap" was replaced by a phenyl-piperidinyl group, which would have better physiochemical properties thanks to the nitrogen-containing piperidine ring. More importantly, the addition of four additional fluorine atoms undoubtedly boosted its resistance to metabolism. The result was a better NS5A inhibitor, pribentasvir. Abbvie combined it with its NS3/4A inhibitor glecaprevir and sold the drug since 2017 with a trade name Mavyret. This pill only has two drugs in it, making prescribing doctors' and pharmacists' jobs easier.[38]

The FDA approved HCV NS5A protein inhibitors are listed below:

- ledipasvir (Harvoni, combo with Sofosbuvir, Gilead) in 2014,
- ombitasvir (a component of Viekira Pak, in combination with NS3/4A inhibitor paritaprevir, dansanovir, and ritovavir Abbvie) in 2014,
- daclatasvir (Daklinza, BMS) in 2015,
- elbasvir (a component of Zepatier, in combination with NS3/4A inhibitor grazoprevir, Merck) in 2016,
- pibrentasivir (a component of Mavyret, in combination with NS3/4A inhibitor glecaprevir, Abbvie) in 2017,
- velpatasvir (a component of Epclusa, in combination with NS5B inhibitor sofosbuvir, Gilead) in 2016, and (a component of Vosevi, in combination with NS3/4A inhibitor voxilaprevir and sofosbuvir, Gilead) in 2017.

3.3.4 Hepatitis C virus NS5B polymerase inhibitors

The HCV NS5B protein is an RNA-dependent RNA polymerase (RdRp) that is responsible for the replication of the viral genome. X-ray crystallography studies revealed that the protein adopts a right-handed three-dimensional structure that is composed of fingers, palm, and thumb subdomains, features that are common among viral polymerases. The Gly317-Asp318-Asp319 catalytic triad found in the palm domain is an *invariable element* in RNA viral polymerase, which means it is highly conserved across all six genotypes. That makes the catalytic triad a perfect target to deliver a pan-genotypic direct-acting antiviral drug with a higher barrier to resistance.

HCV NS5B inhibitors can be categorized into two classes:

(1) Nucleoside inhibitors, compounds that interact with the active site of the polymerase as substrate mimetics and lead to the termination of the replication process; and

(2) Nonnucleoside inhibitors, compounds that inhibit the enzyme's function by interacting with one of several allosteric sites. In general, the former inhibitor class exhibits a superior resistance barrier and pan-genotype inhibitor activity, properties that are sought after in the emerging optimal standard of care for HCV.

3.3.4.1 *Nucleoside/tide inhibitors*

Idenix Pharmaceuticals in Cambridge, Massachusetts was the first to put a nucleoside HCV NS5B polymerase inhibitor in the clinics. Its valopicitabine, a valine prodrug of a nucleoside, advanced to Phase III clinical trials but was discontinued around 2008 due to an unfavorable risk–benefit profile. Its therapeutic index was too small, such that its toxicities outweighed its therapeutic effects.

Hoffmann-La Roche's balapiravir is a triple-ester prodrug of an azido-ribonucleotide. Unfortunately, its development was terminated based on the appearance of unacceptable toxicological effects in clinical trials. Those failures were not unique at all. From 2012 to 2013, more than ten NS5B polymerase inhibitors failed during clinical trials. The drug graveyard was littered with dozens of failed NS5B inhibitors.[39]

Nevertheless, success eventually arrived. It came as a spectacular success in the form of Pharmasett's sofosbuvir (Sovaldi). Pharmasset in Princeton, New Jersey was founded in 1998 by Raymond Schinazi and Dennis Liotta at Emory University. The duo already appeared in Chapter 2 as the inventors of one of the most popular second-generation HIV protease inhibitors, emtricitabine (FTC, Emtriva, Gilead, 2003).

Like many small biotech companies, Pharmasset at first did not know exactly what they wanted to focus on, except for taking advantage of Schinazi's virology prowess and Liotta's experience with nucleoside antiviral drugs. The company dabbled with HIV protease inhibitors, which did not seem to go anywhere. They then worked on hepatitis B drugs, and succeeded with a nucleoside antiviral drug, clevudine (Levovir), which occupies such a tiny market share that it is only sold in South Korea and the Philippines.

Their major and only spectacular victory was their HCV drug, sofosbuvir, as a nucleoside NS5B inhibitor. More impressively, they succeeded where everyone else failed.

What was their "secret sauce"?

Nucleosides and nucleotides are normally the best choice for a clean drug. They tend to have class-related toxicity associated with off-target activity against human DNA and RNA polymerase. They did not have a sterling history of good bioavailability either. Furthermore, nucleosides must be phosphorylated *in vivo* to be active, and most traditional drug design tactics do not really apply, making the discovery a challenging enterprise.

The key to Pharmasset's success is three-fold. (1) They discovered a nucleoside that was very selective against NS5B polymerase. (2) They overcame the low oral bioavailability issue by choosing a prodrug wisely. Finally, (3) they overcame the phosphorylation hurdle using the *kinase bypass* tactic.

Pharmasset came up with PSI-6130, a 2'-α-fluoro-2'-β-methyl nucleoside as a potent and selective inhibitors of HCV NS5B polymerase. PSI-6130 exhibited antiviral activity in cell culture systems and was clinically efficacious. It was a pretty good drug with high potency in an HCV replicon assay. It had a high genetic barrier to resistance and a good safety profile, and it was not cross-resistant with the parent drug of clevudine, a hepatitis B drug. Understandably, nucleoside PSI-6130 demonstrated only modest oral bioavailability in preclinical studies. Pharmasset initially prepared the diester prodrug mericitabine, which is a diisobutyrate of PSI-6130. Mericitabine was superior to the parent drug PSI-6130 and demonstrated pronounced antiviral effects in HCV-infected subjects in clinical trials. Pharmasset codeveloped mericitabine with Roche. Sadly, the drug was not efficacious enough to warrant further clinical trials; thus, it was terminated.

At that stage, Pharmasset isolated PSI-6206, a metabolite of PSI-6130 from the action of the cytidine deaminase. The half-life of the triphosphate of PSI-6206 was significantly longer than that of PSI-6130, indicating that PSI-6206 could be more effectively delivered to cells. Wisely, Pharmasset chose to pursue the metabolite PSI-6206 instead. Detailed metabolism studies conducted in replicons and hepatocytes established that the poor antiviral activity associated PSI-6206 was due to the molecule being a poor substrate for 5'-phosphorylation, the initial metabolic activation step preceding triphosphate formation. These observations prompted the adoption of a *kinase bypass strategy* that relied upon delivering the monophosphate derivative to cells, a tactic that necessitated the use of prodrug elements to mask the highly polar nature of the phosphate moiety, physical properties that limited membrane permeability.

Pharmasset chose the *phosphoramidate* prodrug for PSI-6206. Those *Pro*drugs of nucleo*Tides* are known as ProTides, developed in 1990 by Professor Chris McGuigan at Cardiff University in Wales. Because one molecule of phosphate is already built in the ProTide, there is no need for initial kinase activation to phosphorylate

PSI-6206. That is the reason why the ProTide approach is also known as "kinase bypass" strategy. Such a ProTide can enter the cell via facilitated passive diffusion through the cell membrane. Once inside the cell, the monophosphate nucleoside is released and does what it supposed to do: serve as a viral RNA-replication terminator. ProTide is probably the most successful prodrug strategy applied in the antiviral field.

A normal ProTide has a phosphor atom that is chiral. Therefore, the ProTide prodrug of PSI-6206 is a mixture of two diastereomers. PSI-7977, the single (S)-p diastereomer of the 1:1 mixture of two diastereomers, was more potent of HCV replication. It would later become sofosbuvir. Early clinical trials with PSI-7977 in treatment-naïve genotype-1 subjects suggested considerable promise, with ninety-eight percent of those completing the treatment regimen achieving a sustained virological response defined as undetectable viral load twelve weeks after the end of therapy. It was well tolerated with no virological breakthrough. In ensuing clinical trials, sofosbuvir succeeded beyond everyone's wildest dreams. A combination of sofosbuvir and ribavirin achieved a one hundred percent sustained virologic response in twelve weeks. These astonishing results set the stage for a revolution.[40]

By 2011, even though sofosbuvir was the only genuine asset, the market capitalization of Pharmasset reached $7 billion, the same value of Vertex at the same time! In November, Gilead's CEO and Chairman, John C. Martin, announced that Gilead was buying Pharmasset for about $11.2 billion, a whopping eighty-nine percent premium and one- third of Gilead's market value. Because Pharmasset CEO Schaefer Price owed three percent of the company, he easily made $255 million. One of the founders, Raymond Schinazi, held 4.4 percent of the company and ended up with $440 million from the transaction. Even bench scientists in the eighty-two-employee-company became multimillionaires.[41]

Meanwhile, a similar deal did not work out as well for Bristol-Myers Squibb. In the highly competitive frenzy that was the HCV drug market, BMS decided in early 2012 to buy Inhibitex, a biotech company in Alpharetta, Georgia, for its nucleotide HCV NS5B polymerase inhibitor INX-189 for $2.5 billion. Less than eight months later, clinical trials for the drug INX-189 (BMS-986094) came to a halt due to heart and kidney toxicities. A twenty-five-year-old male patient died of heart failure during the trial and eight others had to be hospitalized. At the end, BMS had also to take an additional $1.8 billion charge for BMS-986094, a key part of its plan to develop an all-oral drug for HCV without interferon. The Motley Fool, a stock advisor website, called the crash and burn event the most epic drug failure and nominated BMS's CEO Lamberto Andreotti for the worst CEO of 2012.

At the end of 2013, sofosbuvir (Sovaldi) was approved by the FDA to combine with ribavirin as a treatment of HCV genotypes 1–4. That was the first HCV treatment without combining with the troublesome interferon. Even though Wall Street cried that it was overpriced for the Pharmasset buyout, Gilead still made out ahead because Sovaldi was a financial windfall. It was famously sold for $1,000 a pill, which drew anger from patients, insurance companies, and even the Congress. HCV at one point was estimated to be a $20 billion market. At Pharmasset, the head of chemistry, Michael Sofia, who joined Pharmasset in 2005, was one of the key contributors to the discovery of sofosbuvir. In 2016, Michael Sofia was awarded the Lasker–DeBakey Clinical Medical Research Award along with Charles M. Rice of the Rockefeller University and Ralf Bartenschlager of the University of Heidelberg for their work that allowed for the development of a therapeutic drug against chronic hepatitis C.[42]

3.3.4.2 *Non-nucleoside HCV NS5B inhibitors*

Non-nucleoside inhibitors are divided into four classes depending on the region of the HCV NS5B protein that they interact with. Consequently, they are referred to as thumb-I, thumb-II, palm-I, and palm-II inhibitors, or alternatively as site I, II, III, and IV inhibitors, respectively. The multiplicity of the biochemical sites for intervention has sustained considerable effort directed toward optimizing a structurally diverse array of inhibitors suitable for development, and several candidates have advanced into clinical trials. One NS5B polymerase inhibitor is on the market in the United States and another in Japan.

Nesbuvir, a palm site II inhibitor was discovered by ViroPharma in Exton, Pennsylvania. It is an allosteric site NS5B polymerase inhibitor that binds proximal to the catalytic triad of the enzyme. It was the first-generation non-nucleoside NS5B inhibitor to demonstrate proof of concept in clinical trials. Nevertheless, its development was terminated due to the appearance of severe hepatotoxicity several weeks into the study. ViroPharma's lomibuvir, a thumb site II inhibitor, was also terminated during it Phase II clinical trials.

Pfizer's filibuvir, a thumb site II inhibitor, was the first nonnucleoside HCV NS5B polymerase inhibitor that went to Phase II clinical trials. Even though filibuvir showed efficacy when combined with pegylated interferon and ribavirin, Pfizer discontinued its development due to strategic reasons in 2013.

Another NS5B polymerase inhibitor, setrobuvir, a palm site I inhibitor, was discovered by Anadys Pharmaceuticals in San Diego. Roche acquired Anadys in 2011 for $230 million. Setrobuvir belonged to the benzothiazidine chemotype class that originated as a high-throughput screening. After Phase IIb clinical trials, Roche terminated the development in July 2015.

Gilead's tegobuvir, a palm site II inhibitor, originated from hits generated from a phenotypic bovine diarrhea virus cellular assay. But the compound was inactive in HCV polymerase biochemical assays. It was found that the active drug was actually the metabolite of tegobuvir generated by CYP450 oxidation. The drug probably had a premature death before reaching the market judging by its disappearance from literature.

Abbvie's dasabuvir is an NS5B polymerase inhibitor of the palm site I class. Its genesis traces back to a dihydrouracil hit from a high-throughput screen. Culmination of their efforts led to the approval of dasabuvir, in combination with the NS5 inhibitor ombitasvir, the NS3 protease inhibitor paritaprevir, and the pharmacokinetic enhancer ritonavir, as a combination drug with a trade name Viekira Pak.[43]

For BMS's NS5B polymerase inhibitors program, chemists began by employing competitors' structures as their starting point, a strategy often known as "patent busting." The compounds that they used were Japan Tobacco's JTK-109 and Merck's MK-3281. BMS chemists at the Wallingford site applied many medicinal chemistry skills and arrived at a pentacyclic indole, which would become beclabuvir, a thumb site I inhibitor, in due course. Interestingly, both JTK-109 and MK-3281 had a cyclohexyl motif, which was crucial to the potency, so it was maintained throughout the drug discovery process. Beclabuvir also had the cyclohexyl motif dangling around the molecule. In 2016, a combination drug with beclabuvir, NS3/4A protease inhibitor asunaprevir and NS5A inhibitor daclatasvir was approved for marketing in Japan with a trade name of Ximency.[44]

HCV NS5B polymerase inhibitors on the market are listed below:

- sofosbuvir (a component of Harvoni, in combination with ledipasvir, Gilead) in 2014,
- dasabuvir (a component of Viekira Pak, in combination with NS3/4A inhibitor paritaprevir, NS5A inhibitor ombitasvir, and ritovavir, Abbvie) in 2014, and
- beclabuvir (a component of Ximency, in combination with daclatasvir and asunaprevir, BMS) in Japan in 2016.

The discovery of HCV drugs that will in time eradicate hepatitis C has been one of the crown jewels of the last decade for drug discovery in general and medicinal chemistry in particular. Scientists continue to invent life-saving medicines to combat today's COVID-19.

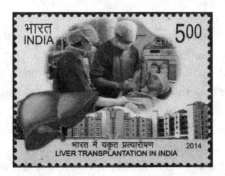

Fig. 3.8 Liver Transplant © India Post

3.4 A Triumph of Modern Medicine

Almost all efficacious and safe HCV direct-acting antiviral agents violate the Lipinski's rule of five. Chris Lipinski, a Pfizer computational chemist came up the rule-of-five in the 1990s. After scrutinizing all the small-molecule drugs on the market, he concluded that for a drug to be bioavailable, or drug-like, a molecule often follows the rules. In short, Lipinski's rule-of-five is an approximate measure of whether the solubility and permeability of the compound exceeds levels for a "typical" drug molecule. An alert is set for the compound if two or more of the following exceed these limits:

- Molecular weight (M) > 500.0
- Clogp (C) > 5.0
- Hydrogen bond acceptors (A) > 10
- Hydrogen bond donor (D) > 5

Otherwise, the compound is marked OK.

The publication of Lipinski's rule-of-five was a watershed event that had a long-lasting impact. Scientists, like all human beings, like shortcuts. If there is a simple rule to follow, life is so much easier. It did not help that Lipinski's rule-of-five played right into the hands of managers who could always wisely point out during a presentation that one or two of the molecules are out of the space of the rule-of-five. Meanwhile, other computational chemists came up with their own rules periodically. On the one hand, most of the molecules come out of big pharma have been rule followers and, therefore, possess better physiochemical properties. On the other hand, there are so many excellent drugs (ten percent) that have been missed due to conforming with the dogma, which greatly hampers the innovation of drug design.

Rules are made to be broken. Many drugs are beyond the rule-of-five. Almost all the HCV direct-acting antiviral agents are beyond Lipinski's rule-of-five drugs. Why? We are not completely sure, except that there are many factors in play. For instance, "chemical chameleon" could play a role. There are molecules, such as cyclosporine A, have extensive polar groups that are not amenable to penetrate the cell membrane. But cyclosporine A forms four intramolecular hydrogen bonds and becomes significantly greasier so that it can go into the cell. Today, drugs beyond the rule-of-five are becoming more and more popular.

In the end, the discovery of hepatitis viruses and effective therapeutics is truly a triumph of modern medicine. HAV was discovered using immune electron microscopy; HBV with the antigen-antibody reaction; and HCV using cloning, which represented the pinnacle of virology.

The discovery of effective hepatitis therapeutics is a triumph of scientists at academia, government agencies, and especially in the pharmaceutical industry. The success is a culmination of advances in pharmacology, medicinal chemistry, computational chemistry, drug metabolism and pharmacokinetics, toxicology, and clinical science. The knowledge gained from the experience will be invaluable for us to conquer the next invisible enemies, such as SARS-Cov-2.

Fig. 3.9 19th World
Transplant Games 2013
© South Africa Post

Most important, the success in the field of hepatitis is a triumph
for patients. With effective vaccines for HAV and HBV, as well as
therapeutic agents as cures for HCV, patients' lives are saves and
their quality of life elevated (see Figure 3.9). Indeed, the success in
the field of hepatitis is a triumph of modern medicine.

4

Influenza

A Perennial Killer

Those who cannot remember the past are condemned to repeat it.

—*George Santayana (1863–1952)*

You know the drill: chill, fever, muscle aches, running nose, and respiratory complaints. You've been there before: it's the seasonal flu, or influenza.

It often happens just before the onset of winter, a time when both the low temperature and low humidity allow the germs spread more easily. The lower the humility, the longer the virus survives. Spreading through aerosol droplets, fomites, or contact with contaminated surfaces, the influenza virus can sometimes cause acute, contagious viral infections of the respiratory tract. The incubation time of the influenza virus is one to three days after one catches the virus with enough viral load. Each year, the world sees about four million cases of severe influenza infection and influenza epidemics are estimated to be responsible for almost 500,000 deaths annually.

Conquest of Invisible Enemies. Jie Jack Li, Oxford University Press. © Oxford University Press 2022.
DOI: 10.1093/oso/9780197609859.003.0004

4.1 Influenza, A Perennial Killer

Influenza is not unique to humans. Many animals, especially domestic animals, carry the influenza virus as well. In fact, wild birds are natural reservoirs for influenza viruses, which obviously do not necessarily need humans to survive. Although most animals carry the influenza viruses, not all of them suffer the symptoms of influenza as humans do. Animals having the flu symptoms include birds, white mice, ferrets, pigs, and even horses. Occasionally, avian flu devastates the poultry industry and swine flu epidemics decimate hog farms. Because swine flu and human flu occur in the vicinity at the same time so often, it strongly suggests that the virus jumped from pigs to humans or *vice versa*. Thus, pigs are recognized as a "mixed vessel" of influenza viruses. In contrast, some animals seem to be oblivious to the influenza viruses. Attempts to infect guinea pigs, rabbits, and monkeys have not resulted in resounding success.

The first account of an influenza epidemic was recorded in 431 BC by Hippocrates. The term influenza, though, was not coined until 1357 AD. It was derived from Italian *influenza di freddo*: influence of the cold. Residents in Florence believed in an unusual conjunction of planets at times of flu epidemics.[1]

Influenza epidemics occur almost annually, sometimes taking on a global scale and turning into pandemics. An epidemic is a local or national outbreak, whereas a pandemic is global, impacting the whole humanity. Twelve pandemics of influenza are believed to have occurred within the last four hundred years. Two Polish medical historians suggested that eleven of them originated from China.[2] That opinion might be biased. History, even medical history, is always tainted by politics, a phenomenon that has never disappeared and never will. For instance, the Jewish people were blamed for plagues in the medieval times and the Chinese people are blamed for the coronavirus today. We are seeing plenty of politics associated with the COVID-19 pandemic.

When influenza kills, it kills the victim in one of the two ways, one is a *violent viral pneumonia* targeting the lungs and the other is *slow bacterial pneumonia*, which also target the lungs after the virus weakens the body's immunological system. Each year, influenza kills more Americans than any other infectious diseases, including AIDS.

Influenza viruses have two types of protuberance on the surface known as *spike proteins*, both glycoproteins. One is hemagglutinin (HA), and the other is neuraminidase (NA). To date, there are eighteen types of hemagglutinins designated as H1–H18 and eleven types of neuraminidases designated as N1–N11. Each virion uses hemagglutinins and neuraminidases in concert to make contact with cells, penetrate the cell membrane, and get into the cell. Once in the host cell, the virus uses the cell's own biochemical machinery to make the next generation of daughter viruses.

Hemagglutinin protein is so named because it causes red cells to clump, known as *agglutination*. Occurring when the *antigen-antibody reaction* takes place, agglutination also lyses the cells; that is, it disintegrates the cell and leaks its content of the pigment and *hemoglobin*. The other spike protein, neuraminidase, breaks up the neuraminic acid remaining on the cell surface and destroys the acid's ability to bind to influenza viruses. Because the neuraminic acid is also known as sialic acid, neuraminidase is sometimes called sialidase. In essence, neuraminidase helps newly made viruses to burst the cell open so that the daughter viruses can get out and go on infecting new host cells. Neuraminidase has proven to be the most fruitful target for antiviral drugs against the influenza viruses.

For a pandemic to occur, the surface spike proteins of the virus must undergo radical changes. In fact, a flu pandemic develops when the influenza virus' hemagglutinin, or neuraminidase, or both undergo mutations. Normally, an RNA virus mutates much faster than a DNA virus. Influenza viruses, along with HIV and coronavirus, happen to be RNA viruses so they mutate faster. To date, H1, H2, and H3, and N1 and N2 have been more frequently found as components of epidemic viruses in humans. New strains are usually produced as the result of the recombination of a human strain as a segmented RNA virus with an avian or swine virus during co-infection, whereby the human strain acquires surface protein genes (among others) from the non-human virus. The new antigen becomes very different from the old one. This process is known as *antigen shift*.

A more substantial antigenic change can occur through *gene reassortment*. When gene reassortment occurs in such a way, the virus undergoes major changes to the hemagglutinin and/or neuraminidase antigens, yet it retains the capacity to cause disease and transmit among humans. These reassortment events may create new pandemic influenza viruses that could cause substantial disease, including deaths, globally.[3] Some pandemics can be exceptionally severe, as evidenced most prominently by the 1918 Spanish influenza pandemic with fatality rates exceeding 2.5 percent and worldwide total death estimates exceeding fifty million.

4.2 A Virus that Changed the World: The 1918 Spanish Flu Pandemic

The 1918 Spanish flu was the deadliest event in human history, ever. Nobody is really sure exactly where the 1918 influenza started, although many believe that it was likely originated from the United States at the tail end of the World War I, the war to end all wars. Some suggested the likely origin of that pandemic was Texas. Others pointed to Kansas. Indeed, one of the earliest occurrences took place at the army cantonment Camp Funston in Haskell County, Kansas where many recruits came down with flu in February 1918. The became the earliest recorded large-scale flu outbreak. Nothing serious. After a three-day fever, most recovered. Later, in San Sebastián, Spain, many locals contracted the flu as well. Nothing alarming either. They recovered after a few days of fever, headaches, and pains. In that Spring, one-third of Madrid was sick with flu. Apparently, even royal palaces could not shield invasion of the virus. Tshe Spanish King, Alfonso XIII, also caught it, prompting newspapers' headline: "The Spanish King Caught the Spanish Flu." The journalists just could not let a good pun go wasted. The plague of 1918 has been called Spanish influenza, or Spanish flu, possibly because Spain was neutral during the Great War. Spanish press was able to report the epidemic freely. In contrast, belligerent countries had severe censorship, keeping the flu outbreaks under the wraps due to military or political considerations.[4]

Fortunately, as summer came, the influenza just miraculously went away. Unfortunately, the virus came back in the fall. And it came back with vengeance.

Like a wildfire, the pandemic ruptured in the fall of 1918 around the world rapidly, violently, and almost simultaneously. It appeared globally almost everywhere at once, even remote, isolated islands were plagued by the flu. The virus became more virulent, more contagious, and more ferocious. Many patients quickly developed severe cyanosis due to lack of oxygen caused by violent viral pneumonia. Sometimes cyanosis was so severe that it extended from their ears to their faces such that it was hard to tell if they were black or white. Even their ability to smell was affected. Similarly, COVID-19 patients lose their olfactory function after infection by the SARS-Cov-2 virus. When patients started having black feet, they were unlikely to survive. Many died within days, and some even just dropped dead while walking! November 11, 1918, the Armistice Day marking the end of the Great War, should have been a day of great joy to the world. But the joy was overshadowed by the ravaging of the second, deadlier wave of influenza.

The public health authorities hastily prepared vaccines against the Spanish flu using the *Pfeiffer's bacillus*. Friedrich J. Pfeiffer, a disciple of Robert Koch, was the scientific director at Berlin's Institute of Infectious Diseases and a general of the German army. In 1892, he isolated a bacterium from the respiratory tracts of flu patients and proposed that it was the cause of influenza. Without any better explanation, that was accepted as dogma and the bacterium was dubbed "Pfeiffer's bacillus" (*Hemophilius influenza*). In truth, that bacterium was merely the secondary infection. Influenza virus weakened the immune system so that the body was prone to the secondary bacterial invasion. The vaccine against *Pfeiffer's bacillus* was not really helpful because it was barking at the wrong tree. Nevertheless, vaccination might have had the psychological effect of calming hysteria of the public.[5]

It was calculated that the Spanish flu was twenty-five-times deadlier than an ordinary flu. The US Navy and Army had about forty percent of their servicemen stricken by the influenza. Overall, twenty-eight percent of Americans caught it. So many Americans died that the average lifespan in the United States fell by 12 years! An average American lived for fifty-one years in 1917, but only thirty-nine years in 1918! No wonder the 1918 Spanish flu is called the Mother of all pandemics. The Spanish flu was one of the worst killers even known in human history, killing more humans than any other disease in a period of similar duration in the history of mankind. MacFarlane Burnet, an eminent Australian virologist and Nobel laureate in 1960, estimated that the death toll was fifty million to one hundred million worldwide with a population of just 1.8 billion at the time (see Figure 4.1).

The number of mortalities underestimates the true horror of the disease because the worst mortality figures fell upon those aged twenty-one to thirty, at the flower of their youth. Indeed, a peculiarity of the 1918 Spanish flu is that it struck mostly the young and healthy populations, such as soldiers ready to be shipped to Europe to fight the Nazis. The phenomenon has been baffling to scientists ever since. Many theories exist. One of them proposed that the more robust young patients were able to incite too many cytokines to encounter the virus invasion. It was the "cytokine storm" that preferentially murdered the healthier victims. Another theory speculates that the previous influenza pandemic some thirty

Fig. 4.1 Jean Macnamara and Frank
Macfarlane Burnet © Australia Post

years prior, from 1889 to 1892, might have something to do with it.
People born before the 1889–1892 period may have contracted influenza and would have had immunity in 1918 when they were over
forty years old. Even people who had the flu in the Spring of 1918
seemed to have immunity against the virus in the fall of that year as
the death rate for those people was significantly lower than people
who did not catch the flu in the Spring.[6]

Among the millions of victims of the Spanish flu pandemic was
Sir William Osler, one of the four founding physicians of the Johns
Hopkins University. He succumbed to influenza at the end of 1918
while visiting Oxford University in England. Another victim was
the father of R. B. Woodward, who would win the Nobel Prize
in 1965 for his outstanding achievements in the art of organic
synthesis.

Deaths alone did not fully convey the misery. The legacy of
the 1918 influenza pandemic included that many survivors still
suffered cardiovascular diseases. Some survivors felt that their
mental states were not the same, even after recovery. President
Woodrow Wilson could be one of them. During the Paris
Conference after WWI, Wilson caught the Spanish flu during
the post-war negotiations. Suddenly and bafflingly, Wilson abandoned all his previous principles after recovering from the flu.

Historians believe that the Paris Peace Treaty helped create the economic hardship, nationalistic reaction, and political chaos that fostered the rise of Adolf Hitler.

Was Wilson's bizarre behavior in Paris a consequence of his influenza?

4.3 Discovery of the Influenza Virus

Due to the tremendous devastation caused by the Spanish flu, intense efforts were underway during and after the pandemic to isolate the etiologic agent that caused the influenza. However, numerous attempts by scientists around the world failed for nearly two decades.

A key breakthrough came when Richard Shope and Paul Lewis, at the Rockefeller Institute in Princeton, discovered the etiology of swine influenza in 1931. Years before, Lewis was the first to prove that a virus caused polio. In 1926, Shope and Lewis studied the swine influenza epidemic in Iowa. From sick animals, they isolated *Pfeiffer's bacillus* with ease. But the bacterium failed to infect other pigs. Shope proposed that the virus was the cause of influenza by linking swine flu to human flu. In 1931, he isolated a *filterable agent* from hogs suffering from swine influenza. The Chamberland porcelain filter that he used had such small pores that only viruses could pass through while larger-sized organisms, such as bacteria, would be trapped behind. Five years later, that same swine flu virus was shown to be a surviving version of the virus that caused the 1918 Spanish flu.

Subsequently, a colleague of Shope's at the Rockefeller Institute, Alphonse Dochez, produced apparent influenza via human nasopharyngeal inoculation and succeeded in cultivating and serially passing a virus in primary chicken embryo cultures, demonstrating that passaged material still produced human disease. But Dochez's results were viewed as ambiguous.

It was not until 1933 that a team of three British scientists succeeded in isolating human influenza virus. They built their discovery on the foundation of Shope's success in isolating swine influenza virus in 1931.

Christopher Andrews, at the National Institute of Medical Research in London, read Shope's papers with great interest and found them compelling. He contacted him immediately. They became fast friends and exchanged their research approaches. In 1933, during a minor outbreak of human influenza in England, Andrews and his colleagues, Wilson Smith and Patrick Laidlaw, discovered the human pathogen. Largely following Shope's methodology, they inoculated ferrets with the filtrate from human influenza patients and successfully caused ferrets to have all the outward symptoms of human influenza including fever and a runny nose. At the time, they already had a supply of ferrets from their studies of the distemper virus. During their research of human influenza virus, Wilson Smith (WS) himself caught the virus and his respiratory tract replicated influenza. That strain of virus, known as the WS strain, is still widely used in virology labs around the world today. Their work on ultrafiltration, isolation, and serial propagation of human influenza virus in ferrets was widely acknowledged as the definitive discovery of human influenza virus. Their discovery contributed greatly to virology, immunology, and molecular biology. They have been given the credit for the discovery because they replaced all earlier work at the level of "textbook-certain proof." In history, credit for discovery may require the kind of clarity that sweeps away earlier contentions. As it so often happens, their monumental discovery required no new technology but rather was made under fortuitous circumstances manipulated by astute observers.

Serologically, the virus that Andrews, Smith, and Laidlaw isolated was termed type A influenza virus. Serologically distinctive influenza B virus was isolated in 1940, and influenza C virus in 1947. Among the three major types of influenza (type D is rare), the worst of them is the type A influenza that has caused epidemics and pandemics. While type B influenza virus causes disease in humans but rarely causes an epidemic, type C influenza virus is innocuous to humans.

The influenza virus is as a single-strand, negative sense, segmented RNA virus that is a member of the *orthomyxovirus* family. Its lipid envelope helps the virus penetrate the cell membrane. The simple influenza virus has only eight genes. The reason why influenza virus primarily infects cells of the lungs is because the lungs have plenty of the enzyme neuraminidase. which the virus needs to split one of its proteins before making daughter viruses.

How about the 1918 Spanish flu virus *per se?*

Due to its lethal impact, it seemed prudent for scientists to study its genome and potentially prepare a vaccine against the deadly virus. Procuring the influenza virus for the 1918 Spanish flu was carried out by Johan Hultin, a retired pathologist in San Francisco, in 1997. In her riveting 1999 book, *Flu, The Story of the Great Influenza Pandemic of 1918 and the Search of the Virus That Caused It*, Gina Kolata, a science reporter for *The New York Times*, described how the 1918 virus samples were physically obtained.[7]

In 1949, Johan Hultin came from Sweden to America for graduate studies in medicine as a visiting graduate student at the University of Iowa. A chance conversation with William Hale, a virologist from Brookhaven National Laboratories, about the 1918 influenza virus inspired Hultin to look for it from tissue samples of the 1918 flu victims in the permafrost of Alaska. The influenza virus is very fragile; it dies after one hour at room temperature, so only a constant, low temperature would have preserved the virus. The Inuit suffered dearly during the 1918 Spanish flu, as many as ninety percent of them were wiped out in some villages. Because many of the victims were buried deep in the permafrost, chances were that the virus could have survived. In 1951, Hultin assembled a small team and traveled to Brevig, Alaska, right across from the Bering Strait. Even though Hultin was able to remove small pieces of frozen lung tissues from the corpses dug out from the grave of the flu victims, he was never able to grow the influenza virus using the fertilized chick embryo technique. That was a disappointing experience for Hultin.

In 1995, Jeffery Taubenberger, at the Armed Force Institute of Pathology, read an interesting article on John Dalton's color blindness. Dalton (1766–1844) was the English scientist who introduced the atomic theory into chemistry. Using a preserved eyeball sample removed after Dalton's death, scientists were able to determine the genes responsible for Dalton's color blindness by employing a revolutionary technique called polymerase chain reaction (PCR). Inspired by this story, Taubenberger decided to sequence the genes of the 1918 flu virus. Fortunately, he was able to locate fixed tissue samples from American Army recruits who died of the Spanish flu. Those samples were pathology lung tissue specimens, preserved in paraffin. Also using the PCR technique, Taubenberger and colleagues elucidated many partial sequences of virus's DNA. The viral genes that they "fished out" of the lung tissue samples included the neuraminidase, the hemagglutinin, the nucleoprotein, the matrix M1 protein, and the M2 ion channel protein. They published their achievement in *Science* in March 1997, which generated a lot of buzz.[8]

One of the keen readers of their *Science* paper was Hultin. Already retired from his pathologist job in 1997, Hultin felt that having the live virus would stitch all the pieces together, so to speak. At his own expense, Hultin traveled to Brevig, Alaska once more. This time, he was able to retrieve frozen, unfixed lung samples from an obese woman's corpse. The body fat might have helped to preserve the integrity of the virus. Even though the RNA quality was not better than in the fixed tissue samples, the amount of frozen lung tissues obtained by Hultin greatly facilitated the completion of the genome of the 1918 pandemic virus by Taubenberger and coworkers. In 2005, after a nine-year effort, the 1918 viral genome was completed, the virus was reconstructed by plasmid-based reverse genetics, and its pathogenicity was first assessed in mice.[9]

Today, new techniques of nucleotide sequencing of viral genome continue to facilitate tracing the lineage of human viruses to provide useful clues about the origin of their genes.

From the genome, we learned that the 1918 flu virus is originated from an avian population, such as chickens, ducks, or geese. The virus that caused the 1918 Spanish flu was named H1N1 subtype. The H1N1 virus mutated when jumping from fowls to humans. Because human bodies had never been exposed to the H1N1 virus, there was no antibody against it. People who survived the epidemic had relatively more robust immunology systems that fought off the virus infection. Some called it the "original antigenic sin."

The virus responsible for the 1957 "Asian influenza," and two million deaths worldwide, was named H2N2 virus. The 1968 "Hong Kong influenza" pandemic demonstrated an antigenic shift from influenza A viruses of the H2N2 subtype to those of H3N2 subtype. Approximately one million people died of the 1968 flu pandemic. The last influenza pandemic was the "swine-origin" H1N1 influenza in 2009.

4.4 Influenza Vaccines

Even though many of us take influenza vaccines every year, they are not very effective due to the antigenic shift of the influenza virus.

4.4.1 History of vaccination

Vaccines work by taking advantage of a human body's own immune system. Once a person is infected by a virus or a bacterium, the body exposed to the antigens from the pathogens starts to marshal specialized white cells and antibodies to bind to the antigens and neutralize their ill effects. Once antibodies have been generated, memory T-cells and antibodies bound to antigens remain in the body. When the same antigen attacks again, the immune system responds significantly faster than the first timer. This is the principle of vaccines.

As early as in the 1500s, the Chinese developed *variolation* to combat smallpox. It was a sophisticated technique that entailed smearing the scabs of smallpox patients' pox into the freshly cut wounds of the person to be inoculated. Variolation was not very effective in staving off smallpox. It was the English country doctor Edward Jenner who pioneered the modern vaccination in 1796 by employing the relatively benign disease of cowpox to produce immunity to human smallpox. What he used was a live animal virus to elicit immunity for humans. Sixty years later, in 1885, Louis Pasteur used live but *attenuated rabies virus* to make a rabies vaccine after he observed that the virulence of the virus was weakened each time it passaged through rabbits. A few years previously, in 1880, Pasteur had already succeeded in preparing a chicken cholera vaccine using *weakened virus* by simply leaving the microbe in an acid culture for a long time. In 1896, one hundred years after Jenner's first success vaccination, Almroth Wright, at the Army Medical College in London, developed a vaccine against typhoid fever using typhoid bacilli killed by heating.[10]

Since then, great strides have been made in vaccines against many deadly infectious diseases caused by both bacteria and viruses. Emil von Behring was awarded the first Nobel Prize for Physiology and Medicine in 1901 for his discovery of the diphtheria vaccine. By injecting diphtheria toxin to animals, such as a horse, and letting the diphtheria bacterium propagate for some days, he then obtained the vaccine from the horse serum containing antibodies (antitoxin) against the diphtheria bacterium. Today, diphtheria has been largely eradicated. Tetanus vaccine using serum antitoxin was invented at the end of 1890s. Vaccines for anthrax, dysentery, pneumonia, and streptococcal infections were gradually discovered and implemented in the early 1900s using the similar serum antitoxin approach.

Some of the most effective vaccines are attenuated or inactivated vaccines, including smallpox, polio, and measles vaccines. Polio vaccines were among the great achievements during the last century. Jonas Salk developed a killed-virus vaccine which had to be given via injection. Albert Sabin developed a superior live attenuated virus vaccine that could be taken orally. Under the leadership of the WHO, smallpox was eradicated in 1980 thanks largely to the global vaccination campaign.

There are at least three types of vaccines:

- Attenuated virus as vaccine,
- Whole killed pathogen, and
- Vaccines only contain the antigen.

Today, making vaccines that contain only the antigen is the most popular approach because there is no chance of having the whole virus, which could potentially cause infection if the virus were not completely killed or attenuated. Modern vaccines are mostly made by genetic engineering. For instance, Merck's quadrivalent human papillomavirus (HPV) vaccine Gardasil is a recombinant vaccine prepared from the major capsid antigen L1 protein, which is produced by inserting its genetic information into a plasmid of ordinary bakers' yeast. It aggregates to form virus-like particles that provide immunogenicity similar to virions but are completely devoid of HPV's oncogenic genome.[11]

We have come a long way since Jenner's first vaccine. Today, two hundred years later, vaccination has controlled fourteen major bacterial and viral diseases, at least in the part of the world, including smallpox, diphtheria, tetanus, yellow fever, pertussis, *Haemophilus influenzae* type B disease, poliomyelitis, measles, mumps, rubella, typhoid, rabies, rotavirus, and hepatitis B.[12]

Regrettably, despite years of effort, we still do not have highly efficacious vaccines against HIV, tuberculosis, malaria, or numerous other widespread pathogens. We still have a long way to go.

4.4.2   History of influenza vaccines

The first attempts to immunize animals against influenza were made in 1935 in the United States by Thomas Francis, head of the US Army Commission, shortly after the influenza virus was first demonstrated to be the cause of influenza. They used formalin-inactivated virus. In 1936, scientists found a way to grow the influenza virus in the large quantities required for vaccine production. Anatoli Smorodintsev, in the Soviet Union, was the first to attempt vaccination in 1936 with a live influenza vaccine that had been passaged about thirty times in eggs. Smorodintsev's live and attenuated influenza vaccine was used in the Soviet Union for more than fifty years and inoculated over one billion subjects. It was still in use in St. Petersburg at the end of the last century.[13]

In the United Kingdom, Christopher Andrews and Wilson Smith, who discovered the human influenza virus, carried out immunization of animals in 1937. Smith used the ferret as an experimental animal to show that prior infection by influenza virus induced immunity to future challenge.

In the 1940s, MacFarlane Burnet, in Australia, developed a relatively simple method for culturing the virus on chick embryos. To grow the virus, at first, a small window is cut on the eggshell. Then a dose of the influenza virus is injected into the egg white by penetrating the membrane. As the egg is incubated at a warmer temperature, the cells begin to become embryos and the virus grows in the embryo's lungs and is secreted back into the amniotic when the embryo exhales. Meanwhile, when the embryo breathes, it draws the amniotic fluid in and out of its lungs. The virus shuttles back and forth between the lungs and the amniotic fluid. As the number of virus cells increase, the amniotic fluid becomes cloudy in a couple of days, brimming with virus to be harvested. Then the virus is killed by formalin. Centrifuging, absorption with alumina, and elution of red blood cells finally afford the vaccine. Vaccines from embryonated eggs dominated the influenza vaccine production more than eighty years. Most inactivated influenza vaccines used today are generated by growing the viruses in embryonated eggs and then breaking up the whole virus with detergents. The viral hemagglutinin protein is purified to serve as the vaccine antigen, although other components of the influenza virus may be present in the final product.

Because the early flu vaccine was made from chicken embryos, people allergic to egg white could not take the flu shots made in such way. Today, influenza hemagglutinin has been produced in insect cells to induce antibodies without the risk of allergy to egg proteins. When influenza virus B was discovered in 1940, the influenza vaccines became bivalent, meaning they were effective against both influenza virus A and B.

America was the first country to immunize its citizens against influenza in 1945 with inactivated influenza vaccines. In 1976, a H5N1 swine flu was expected to occur in the midst of a presidential campaign (see Figure 4.2). President Gerald Ford reacted aggressively and endorsed mass immunization. The CDC ended up vaccinating forty million Americans against the expectant plague which never came. Some vaccine victims sued the US Government for causing the Guillain-Barre disease via the vaccine. "Uncle Sam" ended up paying $100 million to settle the suits outside the courts. Did the blunder cost Ford's defeat in November 1976? We will never know.

Vaccination is the most effective measure for preventing influenza-associated morbidity and mortality. We have come a long way in human history to combat infections and diseases such as influenza using vaccines. Vaccines only contain antigens and thus cannot infect the body like the whole virus does. Most influenza

Fig. 4.2 Domestic animals and influenza
virus © Democratic People's Republic of
Korea Post Service

vaccines use the spike proteins hemagglutinin and neuraminidase
as antigens.

4.4.3 Influenza vaccines today

In the case of inactivated influenza, the objective is to select the
segments coding for hemagglutinin and neuraminidase and to
combine them with segments coding for the internal genes of
viruses that grow well. Thus, one obtains a vaccine that is safe to
handle but still generates functional antibodies against virulent influenza strains.

In the case of live influenza vaccine, the hemagglutinin and neuraminidase RNA segments were reassorted with a previously attenuated cold-adapted virus. The currently available vaccines are
targeted to the hemagglutinin and neuraminidase glycoproteins
of the virus, and thus must be reformulated frequently due to the
circulating virus propensity to mutate at key antigenic sites. Both
hemagglutinin and neuraminidase mutate quickly, which is why
annual flu vaccines are only about sixty percent effective. The influenza virus undergoes frequent antigenic change. When mutations

occur in influenza virus hemagglutinin and neuraminidase surface glycoproteins, they can evade immunity induced by infection to previously circulating strains. This is the basis for annual influenza epidemics and this necessitates the frequent changes in vaccine composition. Today's flu shots contain vaccines against several subtypes of influenza viruses. More recently, reverse genetics has been used to generate the attenuated strains. Antigen shift makes determining which strain of influenza viruses tricky.

There are currently two live attenuated influenza vaccines in use worldwide. One was developed in the former Soviet Union from an attenuated influenza A/Leningrad backbone and has been manufactured and used in Russia for over thirty years. More recently, through a program spearheaded by the WHO, these Russian attenuated seed strains were provided to developing country manufacturers. Serum Institute of India, one of those manufacturers, now has a licensed trivalent live attenuated influenza vaccine for seasonal use. The second vaccine was developed in the United States from an attenuated influenza A/Ann Arbor backbone and has been approved since 2003.

All current vaccines are recommended to contain the selected Influenza A(H1N1) and A(H3N2) strains, plus either one (trivalent) or two (quadrivalent) influenza B viruses. Quadrivalent vaccines were developed to protect against both B lineages currently in circulation in humans, as it has been difficult to accurately predict the predominantly circulating influenza B virus lineage. Further, in many parts of the world, both lineages have co-circulated.

We now have moderately effective influenza vaccines and a standing production capacity based on a reliable viral platform that can be scaled up to deliver vaccine across the nation within about six months. A broader and more durable vaccine, known as the universal influenza vaccine, would be ideal. Much effort has been focused on inventing a universal broadly protective influenza vaccine for seasonal epidemics.

The remaining challenge now is to develop vaccines that can elicit significantly broader protection against antigenically different viruses that can prevent or significantly downregulate viral replication.

4.5 Influenza Drugs

Several drugs are on the market as treatments and prophylaxes against influenza infection. Unfortunately, none of them are efficacious enough to wipe out the calcitrant virus. Annual flu vaccine is still the best means to combat this perennial killer (see Figure 4.3).

4.5.1 M2 inhibitors

Amantadine is an odd compound. It has a fat, greasy adamantane (containing only carbon and hydrogen atoms) bulk attached to a single amine functional group. At the first glance, it does not look

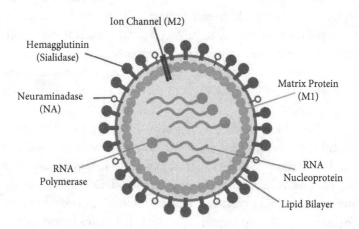

Fig. 4.3 The structure of influenza virus, diagram by Alexandra H. Li

like a drug at all. Indeed, its discovery as an antiviral prophylaxis for influenza was purely fortuitous.

Anecdotal evidence surfaced in the early 1960s regarding the efficacy of amantadine against several strains of the influenza virus. The results were reproduced in tissue culture, chicken embryos, and mice, and it was shown that amantadine either blocked or slowed penetration of the host cells by the influenza virus. In 1963, Jackson and his colleagues at the University of Illinois in Chicago carried out clinical trials and demonstrated that their student volunteers who took amantadine had fewer infections than the placebo group when exposed to a viral challenge of the Asian influenza virus. In 1968, SmithKline & French gained FDA approval of amantadine as a prophylaxis of influenza A subtype H2N2 and sold it under the trade name Symmetrel. In 1976, the FDA further approved amantadine as a prophylaxis for all strains of influenza A. Amantadine was the first specific antiviral treatment of influenza on the market.

Regrettably, clinical use of amantadine has been limited due to its central nervous system (CNS) side effects, lack of activity against influenza B viruses, and a low barrier to the development of resistance. No doubt that amantadine is a "dirty" drug, simultaneously binding to multiple drug targets. In fact, one of its side effects has been taken advantage of to treat Parkinson disease. Moreover, after having been used for six decades, almost all influenza strains must have developed resistance to amantadine by now, potentially due to the extensive overuse of these agents in animals by chicken farmers in China to prevent avian flu. In fact, amantadine is no longer recommended for therapeutic use today.

Back in 1968, a fifty-eight-year old woman with Parkinson's disease reported an improvement in rigidity, tremor, and akinesia while taking amantadine for flu. Yet, her symptoms worsened upon stopping the medication. From a single patient's relief, SmithKline & French began to test on a small population of ten patients, and later on a larger population of 163 patients. They observed

"subjective improvement to excellent subjective and objective improvement" in sixty-six percent of the patients. Even though the
CDC recommended discontinuing using amantadine for the flu
in 2006, it has become an important medicine for early sympto-
matic treatment of Parkinson's disease and an option for treating
dyskinesia.[14]

A follow-up compound for amantadine came much later in
1994 when the FDA approved rimantadine (Flumadine), whose
chemical structure is very similar to that of amantadine. By then,
rimantadine had been in clinical use for many years elsewhere: the
former Soviet Union considered it the drug of choice for treating
influenza A since 1969.

How do amantadine and rimantadine work against influenza
virus? They interfere with the *proton ion* activity of the viral M2
ion channel envelope protein. M2 performs an essential function
during the viral uncoating process, which facilitates the release of
new daughter viruses. Ironically, despite the fact that amantadine
was in use clinically in 1968 based on efficacy and safety data, its
mechanism of action was not elucidated until 1985!

On the surface of the virus, in addition to hemagglutinin and
neuraminidase proteins, it also has matrix M1 proteins and a few
copies of M2 ion channel proteins. The M1 protein is the most
abundant virion protein and it plays an important role in virus
budding. The M2 ion channel mediates the influx of protons into
the virion, thereby destabilizing the association of the influenza
M1 matrix protein with ribonucleoprotein complexes prior to release into the cytoplasm and ribonucleoprotein translocation to
the infected cell nucleus. Amantadine and rimantadine can be
considered as "corks that plug the opening" of the M2 ion channel,
preventing the proton ions from migrating inside the endosome of
the virus and interfering with the viral un-coating inside the cell.[15]

Amantadine and rimantadine are only useful in the treatment of
influenza A infection because only the A strains of the virus have
the M2 ion channel proteins. Influenza B virus strains do not have

an M2 ion channel. Moreover, with ninety-nine percent of circulating viruses resistant to the M2 ion channel inhibitors in recent seasons, amantadine and rimantadine are pretty much useless!

In summary, two influenza viral M2 ion channel inhibitors are listed below:

- Amantadine (Symmetrel, Smith Kline & French) in 1968, and
- Rimantadine (Flumadine, Forest Pharmaceuticals) in 1994.

4.5.2 Influenza virus neuraminidase inhibitors

In contrast to M2 ion channel inhibitors, neuraminidase inhibitors are useful to combat both influenza A and B infections.

Hemagglutinin, one of the two antigenic glycoproteins on the surface of the influenza virus, works at the early stage of the virus lifecycle. It effects attachment of the virus to the host cell through its interaction with surface *sialic acids*, thereby initiating entry. Despite tremendous efforts to find safe and efficacious hemagglutinin inhibitors, none has succeeded thus far.

Neuraminidase works at the tail end of the virus life cycle. Once the virus has replicated, the neuraminidase promotes nascent virus release (budding) from infected cells *via* the cleavage of sialic acid residues at the termini of host cell membrane glycoproteins, allowing the virus babies to spread to uninfected cells. In addition, the removal of sialic acids lowers the membrane viscosity and permits entry of the virus into the epithelial cells, increasing the mobility of influenza within the airway mucosa.

4.5.2.1 *Zanamivir (Relenza)*

Since the early 1940s, sialic acid has been regarded as the primary receptor for influenza virus, although sialic acid itself is not a very potent neuraminidase inhibitor. Early synthetic sialic acid analogs were modestly potent inhibitors of neuraminidase enzyme activity *in vitro* (with only lower μM activity) and virus growth in cell culture. But none of them were active in animal models of influenza virus infection.

To design effective neuraminidase inhibitors, knowing the structure of the enzyme would be immensely useful. Australian scientists made significant contributions to the field.

In 1969, Professor Graeme Laver and colleagues at the Australian National University carried out several expeditions at the Great Barrier Reef in Eastern Australia. From a white-capped noddy tern bird on the North Wets Island, they isolated an influenza A virus, which yielded the "Asian" N2 and N9 neuraminidases. Laver succeeded in crystallizing the N2 neuraminidase from the human H2N2 influenza virus in 1978 and his colleague, Peter Colman, determined its three-dimensional structure by X-ray crystallography. This showed the enzyme to have a conserved catalytic site, which meant that a neuraminidase inhibitor could be designed as an antiviral drug for viruses with N2 neuraminidase. This drug should be effective against all influenza viruses—even those that had not yet appeared in humans.

For the N9 neuraminidase, initial X-ray crystal structures only had a resolution of 3 Å, which had limited value for structure-based drug design. But in the 1970s, Laver was able to grow "the most beautiful large crystals" that diffracted X-ray to 1.9 Å resolution. Because crystals grow bigger and better under weightless conditions, Laver sent the enzyme to grow on an American space shuttle in the condition of microgravity. After the American space shuttle *Challenger* accident in 1986, Laver convinced the Soviet Union to allow his recrystallization experiment to be carried out on their space shuttle *Mir* in 1986. Three months later, the N9 crystals were returned to the Earth and the crystals grown in space were no bigger and of only slightly higher quality than N9 crystals grown on Earth.[16]

In 1992, Joseph Varghese and his coworkers in Australia solved the high-resolution crystal structure of neuraminidase/sialic acid complex. It was a great boon to structure-based drug design that was instrumental to the success of discovering neuraminidase inhibitors.[17]

A very potent neuraminidase inhibitor, DANA, was helpful to obtain good quality cocrystal. DANA stands for 2-deoxy-2-3-dehydro-N-acetylneuraminic acid, the unsaturated sialic acid transition state analog. It was one of the earlier, more potent neuraminidase inhibitors. The crystal structures of neuraminidase/DANA complex revealed their interactions are characterized by both strong charge–charge interactions (salt bridge) and limited hydrophobic contacts. Because DANA was a mimetic of transition state oxonium ion, it was tested close to four-thousand-fold more potent than sialic acid. A transition-state mimetic is an inhibitor that mimics the transition-state structure of the substrate of an enzyme, which, by definition, has the highest energy. Transition-state mimetics also take advantage of the better binding, according to Daniel Koshland's "induced fit" theory. Drugs as transition-state mimetics are often competitive inhibitors because they almost invariably bind to the active sites. Many examples of transition-state mimetics exist in the protease field where inhibitors were designed to mimic the proposed transition-state structure during the cleavage of the substrate. HIV-1 protease inhibitors are more prominent examples of transition-state mimetics.

Mark von Itzstein and his colleagues at Griffith University in Australia designed and synthesized some potent and specific inhibitors of neuraminidase by taking advantage of the more refined crystal structures. Their compounds were evaluated *in vivo* in a mouse model of influenza infection by a team of Glaxo researchers led by Charles Penn and Janet Cameron. von Itzstein's group came up with the first neuraminidase inhibitor that was active *in vivo*, although it was rather weak.

The von Itzstein group was one of the early groups to take advantage of computer-aided drug design. They comprehensively analyzed the cocrystal structure of influenza neuraminidase with

DANA. They identified a region of negative charge in the enzyme proximal to the 4-hydroxyl of DANA. To boost the interaction between the neuraminidase protein and its inhibitors, von Itzstein replaced the 4-hydroxyl group on DANA with a more basic amine group, which would form a stronger salt bridge with the carboxylic acid moiety of Glu-119 on neuraminidase. Indeed, the 4-amino analog bound to neuraminidase more tightly and was tested one-hundred-fold more potent than the prototype DANA. Further optimization focused on exploiting the potential to engage a second proximal glutamate residue (Glu-227), accomplished with the larger and more basic guanidine group to afford zanamivir. Zanamivir exhibited sub-nanomolar potency and a high degree of selectivity for influenza neuraminidases over those of bacterial or mammalian origin.[18]

The highly basic guanidine moiety of zanamivir, while helpful with potency, limits its oral bioavailability and restricts therapy to topical administration, a delivery mode that is efficacious in both murine and ferret models of influenza infection when administered intranasally or by aerosol. For zanamivir, its discovery and early clinical trials were carried out by Biota Holdings, a small Australian biotechnology company. In the United States, Biota licensed zanamivir in 1990 to GlaxoSmithKline (GSK) for clinical development and marketing. Zanamivir was approved for marketing within the United States in July 1999 in the form of an inhaled powder under the trade name of Relenza.[19]

Even though oral inhalation makes Relenza inconvenient to administer, it is less prone to develop drug resistance. Resistant mutants have *not* been observed in immunocompetent persons treated clinically with Relenza.

The successful discovery and development of zanamivir (Relenza) was a feather in the cap of Australian scientists. Some may consider Australia a scientific and medical backwater but, in reality, Australia has produced many eminent scientists who have made significant contributions. Historically in medicine alone,

Australian Sir Howard Florey won the 1945 Nobel Prize for his role in the development of penicillin. Frank MacFarlane Burnet won Nobel laureate in 1960 for his discovery of acquired immunological tolerance. Robin Warren and Barry Marshall won the 2005 Nobel Prize for their discovery of *Helicobacter pylori* and its role in gastric and peptic ulcer disease, just to name a few.

4.5.2.2 Oseltamivir (Tamiflu)

Gilead's orally bioavailable oseltamivir (Tamiflu) is the best-in-class drug among neuraminidase inhibitors.

Gilead Sciences in Foster City, California was founded in 1987 by then twenty-nine-year-old Michael Riordan at Menlo Ventures. He staffed the company with many employees from Bristol–Myers Squibb who did not fare well during the merger of Bristol–Myers and the Squibb Institute in 1989. One key medicinal chemist hire was John C. Martin, who would later become the CEO of Gilead from 1997 to 2016. Martin made many wise business development decisions by buying assets from small companies. One of the best bets that he made was buying Pharmasset with one-third of Gilead's asset. The acquired Solvadi and Harvoni were making so much money for Gilead that some disgruntled consumers mis-spelled Gilead as Greed on purpose. Indeed, $1,000 for one pill is hard to swallow for an average American. But Gilead made enough money that it quickly became one of the top ten pharmaceutical companies in the world, a feat rarely reproduced before and after.

Gilead's team on neuraminidase inhibitors was led by their vice president of medicinal chemistry, Choung U. Kim. Kim had already made a name for himself in the 1970s. While working for Professor E. J. Corey at Harvard as a postdoctoral fellow, Kim co-discovered the Corey–Kim oxidation, a name reaction that has graced many organic chemistry texts. After having worked at Bristol–Myers Squibb at Wallingford for twenty-one years, Kim became one of Gilead's first chemists.

To discover orally active neuraminidase inhibitors, Gilead chose to replace the pyranose core structure on both DANA and zanamivir with a chemically and enzymatically more stable *cyclohexene* scaffold. Using cyclohexene as a suitable bioisostere was a wise choice because it mimics the proposed flat oxonium cation in the transition state of sialic acid cleavage. To fully take advantage of the influenza neuraminidase X-ray structures, Gilead worked closely with Graeme Laver in Australia. To further enhance potency and introduce more drug-like elements anticipated to be compatible with oral bioavailability, a clearly visible hydrophobic pocket in the crystal structure was exploited by the introduction of a lipophilic 3-pentyl moiety in place of the left-handed hydrophilic glycerol side chain of sialic acid and other inhibitor analogs. The X-ray crystal structure revealed that the lipophilic side chains bound to the hydrophobic pocket of the enzyme active site. The increase in potency due to interaction with the hydrophobic pockets was sufficient enough that it was not necessary to incorporate the guanidine group that was critical to the potency of zanamivir. Rather, the lead compound GS-4071 relied upon a less basic primary amine to interact with the region of negative charge in the enzyme, a moiety associated with reduced potency compared to the guanidine found in zanamivir. Later, GS-4071 emerged as one of the most potent influenza neuraminidase inhibitors against both influenza A and B strains. Although considerably less polar than zanamivir, the oral bioavailability of GS-4071 in rats was only five percent. Its ethyl ester prodrug oseltamivir (Tamiflu) had an eighty percent bioavailability in humans. It also proved to be safe and efficacious for the oral treatment and prophylaxis of human influenza infection. Oseltamivir was approved by the FDA in October 1999, just three months after GSK's zanamivir.[20]

After approval of these two neuraminidase inhibitors, Tamiflu has consistently been the more popular than Relenza, obviously due to the convenience of oral dosing when compared to the more unwieldly inhalation treatment required by Relenza. Sales of

Tamiflu soared with the rise in the number of highly pathogenic H5N1 avian influenza cases reported in Asia during 2003 and the subsequent spread to Africa in 2006. The anticipation of a severe avian influenza H5N1 pandemic created a wave of government and institutional stockpiling. With human mortality rates observed in H5N1 infections shockingly exceeding sixty percent, a mutation or reassortment event resulting in efficient human-to-human H5N1 transmission would potentially be a devastating development.[21] Both zanamivir (Relenza) and oseltamivir (Tamiflu) are the anti-viral drugs stockpiled by most countries in the world in pandemic preparedness planning.

The emergence of resistance is always a problem with influenza viruses due to the high mutation rate of the viral polymerase, the selective advantage of resistant viruses in the face of drug treat-ment, and the nature of the influenza season, where viruses are transmitted through human–human contact. For example, during the single influenza season in 2008, the circulating strain of H1N1 influenza evolved to become resistant to oseltamivir, a result of the acquisition of an H274Y mutation: amino acid his-tidine (H) at position 274 mutated to tyrosine (Y). In addition, a growing number of H5N1 cases resistant to this drug have been reported, ostensibly due to resistance in the avian population through overuse on farms. In 2009, a new pandemic strain of flu emerged that was initially sensitive to oseltamivir, but resistance once again developed. Consequently, there remains an unmet medical need for new influenza antivirals and the suggestion that new influenza inhibitors should be combined with the current neuraminidase inhibitors as a means of combating the develop-ment of resistance.[22]

4.5.2.3 *Peramivir (Rapivab)*

The third neuraminidase inhibitor on the market was peramivir (Rapivab). It was developed by BioCryst Pharmaceuticals in Birmingham, Alabama and officially approved by the FDA in 2014, although its emergency use was approved earlier in 2009. Peramivir provides an intravenous (IV) option for hospitalized patients.

Similar to its predecessors, peramivir was developed using a rational design, structure-based approach. BioCryst worked closely with Graeme Laver, and Laver was indeed a busy man, having his fingers on all three of the first marketed neuraminidase inhibitors.

Extensively guided by crystal structures, BioCryst discovered that a cyclopentane ring was a suitable scaffold for a novel class of neuraminidase inhibitors. They routinely used protein crystallography to screen compounds containing mixtures of isomers to identify the active isomer. They believed that it was the first time this technique was extensively used in the successful identification of a new neuraminidase inhibitor. They eventually arrived at peramivir (Rapivab), which takes advantage of the two previous successful drugs. On the one hand, peramivir retains the guanidine motif, mimicking zanamivir, for potency because it occupies the fourth binding pocket replacing the existing water and involves charge-based interactions. On the other hand, its aliphatic 3-pentyl group, mimicking oseltamivir, binds to the hydrophobic pocket of neuraminidase. The interactions between the 3-pentyl group and the hydrophobic pocket were previously not observed in the crystal structure of neuraminidase/sialic acid complex.[23]

It is a common practice for a small biotech company to work with a big pharma company for drug development so that the big pharma with deeper pocket can share the risk and the profit. BioCryst established an alliance with Johnson & Johnson to carry out the clinical trials of their neuraminidase inhibitors. The initial efficacy data in animal models of influenza and preliminary clinical studies suggested oral efficacy with peramivir. However, during phase III clinical prophylaxis trials, peramivir failed to reduce influenza shedding, most likely due to the relatively low blood levels achieved. A parenteral formulation was developed and intravenously administered peramivir were studied in four clinical trials. The 2009 H1N1 swine flu pandemic prompted the

FDA to issue an emergency use authorization (EUA) in the United Sates for peramivir IV use. The FDA official approval was granted in 2014.[24]

4.5.2.4 *Laninamivir (Inavir)*

The most recent neuraminidase entrant is laninamivir, discovered by Daiichi Sankyo in Japan. Laninamivir is the 7-O-methyl derivative of zanamivir and exhibits a similar broad range of activity against most influenza strains. The physical properties of laninamivir are not compatible with oral dosing. A prodrug, laninamivir octanoate, was designed and developed to promote an extended retention time in the lungs, providing a long-acting neuraminidase inhibitor. Preclinical studies have shown that the prodrug is cleaved to the active agent in the lungs, where it resides for extended periods of time, and clinical studies have shown that the active agent exhibits a half-life of three days in humans after a single dose. Laninamivir octanoate, a long-acting neuraminidase inhibitor, showed superior anti-influenza activity after a single administration. In 2010, laninamivir octanoate was approved in Japan as a therapeutic agent for the *treatment* of influenza infection and marketed as Inavir. It was later approved as a prophylaxis of human influenza infection in 2013. Laninamivir octanoate (Inavir), like Relenza, must be given by oral inhalation. The inhaled laninamivir octanoate is converted into its active form, laninamivir, in the lungs where a high concentration persists for a long time. Unlike Relenza, which is given twice daily, Inavir only needs to be given once a week. More importantly, Inavir works for oseltamivir-resistant influenza virus with the H274Y mutation.[25]

There are four influenza virus neuraminidase inhibitors currently on the market:

- zanamivir (Relenza, Biota/GlaxoSmithKline) in July 1999,
- oseltamivir (Tamiflu, Gilead/Hoffmann–La Roche) in October 1999,

- peramivir (Rapivab, BioCryst/Johnson & Johnson) in 2014, and
- laninamivir octanoate (Inavir, Biota/Sankyo) only in Japan, in 2010/2013.

4.5.3 Cap-dependent endonuclease inhibitor

The polymerase complex plays an essential role in the influenza virus replication cycle and, as such, is a major target for the development of small-molecule inhibitors as antiviral drugs. So far, only one drug targeting the viral RNA polymerase has been approved by the FDA, and that is baloxavir marboxil (Xofluza), which was approved by the FDA in 2018.

For an influenza virus, baloxavir marboxil's heterotrimeric RNA-dependent RNA polymerase (RdRp) is composed of three subunits: polymerase acidic protein (PA), polymerase basic proptein-1 (PB1), and polymerase basic proptein-2 (PB2). The RdRp is responsible for replication and transcription of the segmented, single-stranded viral RNA genome in the nucleus of infected cells. Therefore, the viral RNA polymerase is a primary target for antiviral drug development. One particularly attractive approach is interference with the endo-nucleolytic "cap-snatching" reaction by the RdRp subunit PA, more precisely by inhibiting its metal-dependent catalytic activity which resides in the N-terminal part of PA. The PA unit contains two magnesium divalent ions.

Back in 2014, an RNA synthesis inhibitor, favipiravir (Avigan), was approved in Japan although the indication was limited for the treatment of novel influenza viruses unresponsive to other agents. Favipiravir was one of the closely watched antiviral drugs tested against the SARS-CoV-2 virus in 2020.

Endonuclease as a part of the polymerase acidic protein (PA) is crucial to viral replication and is thus an excellent target for antiviral drugs. However, initial attempts by pharmaceutical companies failed, probably due to the failure of identifying potent compounds even in the pre-clinical stages. It was not until 2009 when the structure of PA was determined by crystallography, which revealed that the viral endonuclease function resides in the N-terminus of the PA subunit.

Shionogi Inc. in Japan was inspired by the discovery of an HIV integrase inhibitor, dolutegravir. In 1999, Shionogi discovered a diketoacid derivative that specifically bound to the HIV integrase. The diketoacid moiety, as a phosphate isostere, inhibited the integrase functions by binding to the two magnesium divalent ions involved in catalysis by the HIV integrase. GSK took advantage of Shionogi's diketoacid and used it as a starting point to discover dolutegravir (Trivicay), which was approved by the FDA in 2013.

Circling back and building upon the key structural feature of dolutegravir, Shionogi also built metal-chelating functionalities into their drugs because influenza virus PA endonuclease also has two magnesium ions! Shionogi eventually discovered baloxavir, also known as baloxavir acid. Unlike neuraminidase inhibitors that impair viral release from infected host cells, baloxavir blocks influenza virus proliferation by inhibiting the initiation of mRNA synthesis. To increase the oral bioavailability, its prodrug, baloxavir marboxil, was prepared and found to be orally bioavailable.

As a selective inhibitor of the influenza virus cap-dependent endonuclease, baloxavir marboxil (Xofluza) was approved in Japan and in the United States in 2018 for the treatment of influenza A or B virus infections. Baloxavir marboxil is advantageous because it can be given orally. In addition, because of the lack of cross-resistance with other antiviral medications, it may be used as an alternative for the treatment of viruses resistant to other medications. It can also be used as a combination therapy with neuraminidase inhibitors. Although baloxavir marboxil does not have a significant clinical advantage over oseltamivir, as measured by time to alleviate symptoms, its ability to reduce viral loads within twenty-four hours of treatment makes it superior in preventing virus spread.[26]

In summary, only one RNA polymerase inhibitor as a treatment of influenza has been approved by the FDA thus far:

- Baloxavir marboxil (Xofluza, Shionogi/Roche) in 2018.

5

Coronaviruses

Nature only answers when she is questioned.

—*Francis Bacon*

Before 2020, the word *coronavirus* was familiar only to the virologists. Today, coronavirus is in our daily vernacular, uttered by everyone around the world (see Figure 5.1).

In December 2019, some patrons who visited a seafood and animal market in Wuhan, China, began to show pneumonia-like symptoms. But unlike normal pneumonia, some patients became gravely ill and died at an alarmingly high rate. China's immediate, aggressive response may have delayed the global spread of the current outbreak. By the end of January 2020, the world began to know coronavirus disease-2019 (COVID-19) that is caused by a virus called severe acute respiratory syndrome beta-coronavirus-2 (SARS-CoV-2). The virus spread like a wildfire to twenty-nine countries and many regions around the world and, on March 11, 2020, the World Health Organization (WHO) officially declared it a global pandemic. Even at the time of writing in June 2021, the virus is still rampaging, transmitting, mutating, and killing in almost every country around the world, although at this time the pandemic is largely under control in China and the United States.

Conquest of Invisible Enemies. Jie Jack Li, Oxford University Press. © Oxford University Press 2022.
DOI: 10.1093/oso/9780197609859.003.0005

Fig. 5.1 COVID-19 © Swiss Post

5.1   Coronavirus

5.1.1   What is a coronavirus?

The coronavirus is so named because, when observed by an electron microscope, many spike proteins are seen on the surface of the virus particles. The spikes look like a king's crown (*corona* in Latin). Four main proteins are encoded by the coronaviral genome: spike (S), envelope (E), membrane (M), and nucleocapsid (N). Each protein plays an individual role in the structure of the viral particle; however, they are also involved in other functions of the replication cycle.

Coronavirus was initially recognized in the 1960s when it was found that animal coronavirus-associated diseases, such as feline infectious peritonitis, were part of a group of related enveloped positive-strand RNA viruses. Like all other RNA viruses, coronaviruses mutate frequently and evolve in vast animal reservoirs. In 1975, The International Committee on Taxonomy of Viruses approved a separate virus family, *Coronaviridae*. Prior to the twenty-first century coronaviruses were considered pathogens of great relevance only in veterinary medicine but with a reduced impact on human health. Indeed, the overwhelming majority of coronaviruses pose no threat to humans. Coronavirus is estimated to cause fifteen to thirty percent of all colds. It infects epithelial cells, just like the influenza virus. Typically, coronaviruses cause mild to severe respiratory and intestinal infections in mammals, including humans.

Nevertheless, recombination events, natural selection, and genetic drift permit particular virulent coronaviruses jump to human hosts and subsequently acquire the capacity for efficient person-to-person spread. The world really began to take coronaviruses seriously in 2002 with the outbreaks of severe acute respiratory syndrome (SARS) and, in 2012, with outberaks of the Middle East respiratory syndrome (MERS). Both SARS and MERS were highly pathogenic, and led to communicable outbreaks and fatal respiratory diseases.[1]

5.1.2 SARS and MERS

In 2002, a pneumonia-like disease emerged in Guangdong Province, China, causing severe lung disease. It was caused by a coronavirus: later named as severe acute respiratory syndrome coronavirus (SARS-CoV) (see Figure 5.2). It rapidly spread to twenty-nine countries and many regions around the world. The case numbers reached 8,098, with 774 fatalities by the end of the epidemic in July 2003, which translated to a 9.6 percent fatality rate. Mercifully, the virus did not transmit readily so the damage was limited. Aggressive public health intervention measures managed to contain a potential pandemic. Indeed, the high speed of the scientific response in understanding this new viral disease was unmatched and contributed to the success of SARS containment.

Because the virus that causes SARS shares eighty-eight to ninety-five percent of its sequence homology with the virus in civet cats, it was suggested that bats were probably the natural reservoir of a close ancestor of SARS-CoV, and that palm civets were possible

Fig. 5.2 SARS © China Post

intermediate hosts. SARS-CoV infects human alveolar type II cells. After a virion enters the blood stream, the spikes bind to receptors on the host cell's surface. The receptor is angiotensin converting enzyme-2 (ACE2).

In 2012, Middle East respiratory syndrome (MERS) was first reported in Saudi Arabia and South Korea. The virus, MERS-CoV, caused 2,494 laboratory-confirmed cases of infection and 858 deaths across twenty-seven countries, registering a stunning thirty-four percent fatality rate! It was suspected that the virus was initially transmitted from bats to camels. Intriguingly, the virus enters the cell through the dipeptidyl-peptidase 4 (DPP-4, also known as CD26). This is fascinating because DPP-4 inhibitors are popular diabetes medicines.[2]

5.1.3 SARS-CoV-2

SARS-CoV-2 is extraordinarily contagious and has been fast-spreading across the world. Compared to the previously emerged

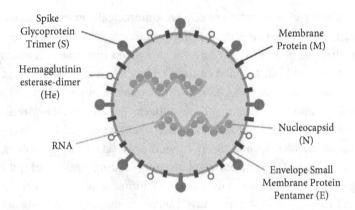

Spike
Glycoprotein
Trimer (S)

Hemagglutinin
esterase-dimer
(He)

RNA

Membrane
Protein (M)

Nucleocapsid
(N)

Envelope Small
Membrane Protein
Pentamer (E)

Fig. 5.3 SARS-2-CoV-2, diagram by Alexandra H. Li

SARS and MERS coronavirus, SARS-CoV-2 causes an unprecedented threat on global health and tremendous socio-economic
consequences.

Just like all other coronaviruses, SARS-CoV-2 is a positive-sense,
single-stranded RNA (ssRNA) virus surrounded by an envelope.
It also encodes four main proteins: spike (S), envelope (E), membrane (M), and nucleocapsid (N). Like its predecessor SARS-CoV,
SARS-CoV-2 spike proteins bind the ACE2 receptors before entering the host cells. Sequences of SARS-CoV-2 show eighty-two
percent homology with SARS-CoV. One of the major reasons why
the COVID-19 pandemic has been so devastating is because the
virus spreads rapidly and efficiently. A conference held by Biogen
involving many biotech companies was attended by about one
hundred high-level managers in February 2020. That conference
turned out to be a "super spreader" event. By end of 2020, scientists
discovered that 245,000 people tested positive from the same strain
of the virus disseminated from that one-day conference.

Once the pandemic broke out in 2020, serologic testing became of paramount importance. There are two popular tests of the
coronaviruses: one is the antibody test and the other is the nucleic
acid test.

Once a virus enters our body, we automatically marshal our autoimmune system to stimulate the production of antibodies to neutralize the pathogen. The area on an antigen that makes contacts with an antibody is called epitope, which serves as a marker of an immune response to SARS-CoV-2 infection. Many serologic tests using antibodies against specific antigens are commercially available. The enzyme-linked immunosorbent assay (ELISA) technique enables identification of IgM and IgG immunoglobins, nucleocapsid (N) antigens, and receptor-binding domains of spike proteins. These tests only reflect an immune response without indicating the host immunity. Therefore, the antibody tests are quick and dirty, and not completely reliable.

On the other hand, the nucleic acid test takes advantage of the polymerase chain reaction (PCR) techniques. It takes longer to get a result, but it is a more reliable test, especially the real-time reverse transcriptase-PCR (RT-PCR) technique. This technique targets the SARS-CoV-2 S, N, E, and RdRp viral genes. The nucleic acid test is more precise and more definitive.

Because of the pandemic, for the first time in decades cardiovascular deaths were no longer the disease that killed most Americans. Sadly, COVID-19 became the number one killer that has claimed more than one million American lives, and over six million lives globally, to date.

5.2 COVID-19 Vaccines

In the face of the COVID-19 pandemic, science came to rescue. In less than one year, three vaccines gained Emergency Use Authorization (EUA) from the FDA in the United States. They are mRNA vaccines developed by Pfizer (Comirnaty) and Moderna (Spikevax), respectively, and Johnson & Johnson's viral vector vaccine (Jcovden).

5.2.1 Pfizer/BioNTech's mRNA vaccine

On November 9, 2020, Pfizer and BioNTech announced that their vaccine candidate against COVID-19 had achieved success in first interim analysis from the phase III study. Their mRNA-based vaccine, PF-7302048 (BNT162), was found to be ninety percent effective against the SARS-CoV-2 virus with a reasonable safety profile. This was on par with the smallpox vaccine that eradicated the infection. In contrast, our annual flu vaccines are only forty to seventy percent effective against the influenza viruses. The Pfizer CEO hailed the vaccine as one of the greatest medical breakthroughs in a hundred years. Pfizer gained approval from the FDA for EUA on December 15, 2020. One week later, Moderna's mRNA COVID-19 vaccine was also granted EUA. This is remarkable and extraordinary because those vaccines started clinical trials only six months before.

Back on January 10, 2020, Dr. Yong-Zhen Zhang at Fudan University in China and his collaborators published the genome sequence of SARS-CoV-2, a map of how the virus is composed.[3] It was posted free online so that scientists around the world could take advantage of the genetic information to make vaccines. A small biotech in Germany, BioNTech, designed and prepared four mRNA vaccine candidates, which could generate the spike protein antigens. Pfizer then licensed the vaccines from BioNTech and spent one billion dollars developing them. They winnowed down four options to one best candidate.

The principle of making a vaccine is to harness the body's innate and adaptive immunology. Once the body encounters an antigen, our immunology system generates antibodies to kill the antigen. The trick is generating the antibodies without actually replicating the virus or other adverse effects. For the SARS-CoV-2 virus, if we inject a vaccine that only contains the surface spike glycoproteins, it will not replicate the virus because the vaccine does not contain

DNA coded with the genetic information of the virus. This was BioNTech and Pfizer's approach for their RNA-based vaccine, which encodes for the receptor binding region of the spike glycoprotein. The mRNA vaccine can take advantage of the host cell translational machinery to produce the antigen proteins and launch an adaptive immune response. Unlike attenuated live vaccines, RNA-based vaccines do not carry the risks associated with infection and may be given to people who cannot be administered a live virus vaccine, such as pregnant women and immunocompromised persons.

Regarding genetic information flow, Francis Crick, the co-discoverer of the DNA double-helix structure, advanced the central dogma for the flow of genetic information in the late 1950s. Central dogma describes the flow of genetic information in a two-step process of transcription and translation in a unidirectional vector: DNA → RNA → Protein. Basically, mRNA is used to generate proteins that resemble the SARS-CoV-2 virus' surface spike glycoproteins.

mRNA stands for *m*essenger *r*ibo*n*ucleic acid. As shown below, RNA is a polymer of nucleotide monomers as building blocks. This is why an RNA-based vaccine is sometimes known as a nucleotide-based vaccine. The development of an RNA-based vaccine encoding a viral antigen, which is then expressed by the vaccine recipient as a protein capable of eliciting protective immune responses, provides significant advantages over more traditional vaccine approaches. In short, mRNA vaccines have three prominent advantages over traditional vaccines: safety, efficacy, and speed.

The concept of exploiting mRNA as a novel therapeutic or a vaccine gained greater recognition in 1989 when it was demonstrated that mRNA packaged within a liposomal nanoparticle could successfully transfect mRNA into a variety of eukaryotic cells. Even injection of "naked," unprotected mRNA into the muscle of mice resulted in expression of the encoded protein for a couple of days! Moving forward, mRNA emerged as an attractive source of antigens. In comparison to protein-based vaccines, antigen-encoded mRNA vaccines have a more profound antibody response, and they result in a more durable protection. However, "naked," unprotected mRNA vaccines do not survive long enough. Thus, mRNA vaccine delivery has been a constant struggle. It has been found that lipid nanoparticle (LNP) is an ideal vehicle for delivering the RNA vaccine. Nanotechnology has played a key role in the success of the vaccine. Nanoparticles and viruses have similar sizes; therefore, nanoparticles can enter cells to enable expression of antigens from delivered nucleic acids, such as mRNA, or to directly target immune cells for delivery of antigen. The Pfizer and Moderna vaccines encapsulate the mRNA vaccines within the lipid nanoparticles whereas CanSino (China) and AstraZeneca (UK) incorporate antigen-encoding sequences within the DNA carried by antibodies.

RNA-based vaccines are manufactured via a cell-free *in vitro* transcription process, which allows an easy and rapid production and the prospect of producing high numbers of vaccination doses within a shorter time-period than achieved with traditional vaccine approaches. This capability is pivotal to enable the most effective response in outbreak scenarios. Therefore, it is not surprising that mRNA-based vaccines crossed the finish line ahead of all other traditional approaches in developing vaccines against COVID-19. Clinical trials demonstrated that PF-7302048 (BNT162) elicited human antibodies and T-cell response. Statistically, the vaccine is ninety-seven percent effective against the SARS-CoV-2 virus.

However, the vaccine must be stored at –70°C. Not many facilities have such an ability—your household refrigerator is not going to cut it. So, there is also a cold supply-chain challenge involved in distributing the vaccine to the whole world and in the hopes of achieving herd immunity.

Even though mRNA vaccines have been studied for over thirty years, no such vaccine has been approved by the FDA before. The Pfizer vaccine submitted material to the FDA for a Biologics License Application in May 2021, and it was approved by the FDA as the first-in-class vaccine.

5.2.2 Johnson & Johnson's viral vector vaccine

On February 28, 2021, the FDA granted EUA to Johnson & Johnson's viral vector vaccine (Ad26.COV2.S) as a prophylactic agent against COVID-19. It became the third vaccine to be available in the United States. Because this vaccine requires only one shot (as opposed to the two-dose schedule of the Pfizer and Moderna vaccines) and because it does not require stringent cold storage, it has some advantages over the RNA-based vaccines.

Nearly every COVID-19 vaccine shares the same objective: preventing interactions between the spike (S) protein on the virus and ACE2 on the surface of host cell. Each SARS-CoV-2 virus has about two hundred spike glycoproteins, which are mushroom-like and which study the spherical surface of the virus. The receptor-binding domain (RBD) sits on top of the spike, where it makes direct contact with ACE2. Once the spike protein grabs hold of ACE2, it undergoes a dramatic transformation. It pulls the virus and host cell so close together that their membranes fuse and this allows the virus penetrate the host cell membrane and enter into the cytoplasm. The process is called tropism, which is the first step of an infection. Once inside the cell, the coronavirus begins replicating by taking advantage of the host cell's own biochemical machinery to make the next generation of virus. Consequently, a virus' spike protein is critical for its viral replication cycle and is the primary target of our neutralizing antibodies. The antibodies, in turn, latch onto invading viruses and prevent them from entering our cells.

Five classes of COVID vaccines are being pursued, all of which contain or make the spike protein:

- attenuated and inactivated virus vaccines;
- mRNA vaccines;
- viral vector vaccines;
- subunit protein vaccines; and
- DNA plasmid vaccines.

Conventional methods of making vaccines using a live-attenuated virus (weakened) or a whole-inactivated (killed) virus are tried-and-true, but they take a long time to develop. They stimulate our B-cells to make antibodies. Sinovac and Sinopharm in China developed inactivated virus vaccines for COVID-19. Like DNA vaccines, inactivated vaccines often need two immunizations, as do both mRNA vaccines by Pfizer and Moderna. In contrast, some viral vector vaccines can induce robust and durable neutralizing antibody responses after a single immunization.[4]

The four frontrunners for making viral vector vaccines are China's CanSino Biologics (CanSinoBIO), Oxford University in collaboration with AstraZeneca, Russia's Gamaleya Institute, and Johnson & Johnson.

5.2.2.1 What is a vector?

Also known as a recipient or a vehicle, a vector is a more innocuous bacterium or a virus (e.g., adenovirus) into which the genetic sequence of the antigen (e.g., spike protein of SARS-CoV-2) is inserted.

5.2.2.2 What is a viral vector vaccine?

Also known as a vectored vaccine, it is a vaccine that uses a vector to deliver genes encoding the antigen as the genetic payload into host-cell nuclei to elicit an immune response.

Viral vector vaccines are considered subunit vaccines because they only make antigens, not the whole virus. They use another more innocuous virus such as adenovirus, poxvirus, lentivirus, vesicular stomatitis virus, herpes virus, and measles virus to carry the genetic instructions to make the spike protein. Current coronavirus vaccines all use adenoviruses for very good reasons. Known to cause mild cold or flu-like symptoms, adenoviruses were first isolated in the 1950s from adenoids that had been surgically removed. As non-enveloped icosahedral double-stranded DNA (dsDNA) viruses, adenoviruses are considered large viruses and are amenable to easy manipulation. They are also very good at getting into cells. Adenoviruses have transitioned from initially being used as tools for gene replacement therapy in the past to being *bona fide* vaccine delivery vehicles. They are attractive vaccine vectors due to their high insert capacity and proven immunogenicity to induce both innate and adaptive immune responses in mammalian hosts.

To construct a viral vector vaccine for COVID-19, the genes encoding the prefusion-stabilized spike protein from SARS-CoV-2 are identified from the virus genome and prepared. Meanwhile, an adenovirus vector is chosen and genetically engineered so it cannot replicate. The modified adenovirus as the carrier virus is fused with the spike protein DNA as the genetic cargo to make the vaccine. This is why a viral vector vaccine is also known as a recombinant vector vaccine. When the patient is given the vaccine, the recombinant adenovirus vector infects the host cell and pushes the spike DNA into the cell nucleus. Subsequently, the gene for the coronavirus spike proteins are read by the cell and copied into a messenger RNA (mRNA). mRNA then tells the cells to make the natural SARS-CoV-2 spike protein using host cell's own biochemical machinery. Some of the spike proteins produced by the vaccinated cell form spikes that migrate to its surface and stick out their tips, giving rise to an *antigen-presenting cell*. Adenovirus vector vaccines are the best of all vaccines at inducing a helper T-cell response.

CanSinoBIO uses replication-defective adenovirus serotype 5 (Ad5) as the viral vector to make their COVID-19 vaccine (Ad5-nCoV). Ad5, which ubiquitously infects humans and causes the common cold, was one of the first adenoviral vectors to be used in the 1980s. Researchers stripped Ad5 of the early E1 gene, rendering it replication-incompetent and inserted those genes into genetically engineered cell lines.

Founded by former Sanofi vaccine developers, CanSinoBIO developed an Ad5-based vaccine for Ebola during the 2014 outbreak, which was approved in 2017 in China for military use. Taking advantage of their expertise with the Ad5 vector, CanSinoBIO quickly developed their COVID-19 vaccine using Ad5 as the viral vector. In March 2020, CanSinoBIO became the first company in the world to begin a clinical trial of a COVID-19 vaccine. Ad5-nCoV (tradenamed Convidecia) has been approved for immunization in several countries.

The major downside to the human adenovirus vectors, such as Ad5, is pre-existing immunity against the vector itself, which could destroy the vehicle and blunt the vaccine's effectiveness. Ad5 circulates widely, causing the common cold and some people, especially the elderly, harbor antibodies that will target the vaccine, making it ineffective. In the United States, about forty percent of Americans are seropositive for Ad5. To overcome such a shortcoming, Oxford/AstraZeneca picked a chimpanzee adenovirus and Johnson & Johnson chose to use adenovirus serotype 26 (Ad26), which has a lower prevalence in humans.

Russia's Sputnik V vaccine (rAD26-S/rAd5-S) starts with a shot of Ad26 vector followed by a booster with Ad5 vector, both of which carry the gene for the spike protein of SARS-CoV-2. The Sputnik-V is reported to be ninety-two percent effective. It was reported that Russian President Putin's daughter had a fever after taking the vaccine. This was probably because vector-based vaccines stimulate a strong immune response, especially for the booster shot.

As alluded to earlier, the Oxford/AstraZeneca vaccine (ChAdOx1 nCoV-19, trade-named Vaxzevria) uses a chimpanzee adenovirus as a vector that closely resembled human Ad5. Chimpanzee adenoviruses have sero-prevalence of below ten percent in most human populations with slightly increased rates in Sub-Sahara Africa. Back in 2012, the Oxford group developed its own chimpanzee-derived vector, dubbed ChAdOx1, based on an adenovirus discovered in chimpanzee feces. They successfully developed a vaccine against the Ebola virus using the chimpanzee adenovirus vector. To make the COVID-19 vaccine, the chimpanzee adenoviral genome is modified to remove viral replicating genes, and the genetic material of the SARS-CoV-2 spike protein is then constructed. This way, the viral vector cannot replicate or cause disease; instead, it acts as a vehicle to deliver the DNA encoding the spike protein. The reason for using a non-human adenovirus is because most people may have been exposed to human adenovirus and have immunity against it.[5]

As soon as they saw the genetic sequence published online by Yong-Zhen Zhang and his colleagues, Janssen Vaccines & Prevention BV in Leiden, The Netherlands, a division of Johnson & Johnson, began independently working on adenoviral vector vaccines for COVID-19, as did several other groups around the world.

To make the COVID-19 vaccine, Janssen scientists chose human adenovirus serotype 26 (Ad26) as the viral vector. Ad26 is a relatively rare virus that causes mild colds but is very effective at invading human cells. With only ten to twenty percent of people in Asia and Europe having immunity, Ad26 is less prevalent in humans than Ad5, although Ad26 is more prevalent in the Sub-Sahara region. Janssen had already done decades of research on adenovirus-based vaccines in general and the Ad26 platform in particular. In July 2020, their Ad26 viral vector vaccine for Ebola (Zabdeno) was approved by the European Medicines Agency (EMA), which marked the first commercial adenoviral vector vaccine proven to prevent a disease in humans. Furthermore, the availability of industrialized and scalable manufacturing processes makes the Ad26 viral vector an attractive platform for vaccine development.

To make the vector, the Janssen researchers disabled the Ad26 virus by deleting its E1 region so that it could only invade cells but not multiply in them (replicating incompetent). They subsequently fused the Ad26 virus with the genetic instruction in the form of double-stranded DNA to make a prefusion-stabilized SARS-CoV-2 spike protein. The vaccine Ad26.COV2.S (trade-named Jcovden) is a replication defective vector. Because Ad26 cannot reproduce itself, a higher dose is needed to be effective.[6]

Like other adenovirus vector vaccines, the Johnson & Johnson vaccine elicits neutralizing antibodies to bind to the spike protein in a manner that prevents the virus from infecting our cells. It also induces T-cells to clear cells infected by the virus. In addition, the immune system also contains special cells called memory B-cells and memory T-cells that might retain information about the coronavirus for years or even decades.

Adenovirus-based vaccines for COVID-19 are more robust than the mRNA vaccines from Pfizer and Moderna. DNA is not as fragile as RNA because the adenovirus' tough protein coat helps protect the genetic material inside. The Johnson & Johnson vaccine can be refrigerated for up to three months at 2–8°C. More importantly, adenovirus vaccines are cheaper, costing about $2.50 per dose, whereas mRNA vaccines cost about $17 per dose, seven-fold more expensive.

In terms of efficacy, Johnson & Johnson's vaccine is underwhelming in comparison to the two mRNA vaccines by Pfizer and Moderna, which have an effective rate of about ninety-five percent. Johnson & Johnson's vaccine has an overall sixty-six percent efficacy rate worldwide, even though it is seventy-two percent effective for patients in the United States and eighty-five percent effective in preventing severe cases. However, there are many advantages to the vaccine. There are no hospitalizations or deaths for patients who took the Johnson & Johnson vaccine in clinical trials, and the vaccine needs only one shot instead of two. More importantly, the Johnson & Johnson vaccine is inexpensive; therefore, it is easy to make billions of doses. As mentioned, it can be stored in normal refrigerators rather than requiring special storage like the mRNA vaccines.

Jonas Salk famously said: "What had the most profound effect was the freedom from fear." With the arrival of Johnson & Johnson's viral vector vaccine, the third COVID vaccine, we will have the freedom from fear of the invisible enemy as we achieve herd immunity in the summer of 2021!

5.3 COVID-19 Drugs

Only one antiviral drug, remdesivir, has been approved to treat COVID-19 patients. Dexamethasone, a steroid, has been remarkable in saving the lives of severely ill COVID-19 patients.

5.3.1 Remdesivir (Veklury)

Gilead's antiviral drug remdesivir (Veklury) was approved by the FDA on May 1, 2020 for emergency use in COVID-19 patients after a record one-day review. It took the Japanese government much longer—seven days—before granting its regulatory approval.

One wonders: how was remdesivir discovered? Its genesis traces back to the first antiviral drug: idoxuridine. Inspired by George H. Hitchings (Nobel laureate in 1988 for Physiology and Medicines), who started to systemically investigate purine and pyrimidine analogs as potential drugs, Dr. William H. Prusoff at Yale discovered idoxuridine (IdU) as the first small-molecule antiviral drug in 1959. He is now known as the godfather of modern antiviral chemotherapy. Although IdU is too toxic to be given systemically, it is applied topically to treat eye and skin infections caused by herpes simplex virus (HSV). While its mechanism of action (MoA) is not completely elucidated, it is most likely phosphorylated first by kinases in both virus and normal cells to the corresponding nucleoside monophosphate (when a phosphate is attached to a nucleoside, it becomes a nucleotide), nucleotide diphosphate, and nucleoside triphosphate sequentially. Nucleoside triphosphate is the active drug with two fates. On the one hand, when interacting with viral DNA polymerase, it terminates DNA replication and exerts antiviral activities. On the other hand, when interacting with cellular DNA polymerases, cytotoxicity, mitochondrial toxicity, and antitumor activity ensues.

The emergence of IdU opened a floodgate of ribonucleoside antiviral drugs. It was followed by trifluorothymidine (Viroptic), ethyldeoxyuridine, bromovinyldeoxyuridine, and more recently, telbivudine (Tyzeka), a synthetic thymidine nucleoside analog put on the market by Novartis in 2006.

Gertrude "Trudy" Elion, who shared the 1988 Nobel Prize with George H. Hitchings, led a team at Wellcome Research Institute that discovered acyclovir (Zovirax) for the treatment of HSV. One of the Wellcome chemists, Howard Schaeffer, established that the intact sugar ring of compounds (such as guanosine) was not essential for binding to enzymes needed for DNA synthesis. Cutting off the diol fragment of the sugar fragment led to the discovery of acyclovir, which became one of the most successful antiviral drugs at the time. Scientists' creativity is boundless!

Following the discovery of acyclovir, its "me-too" drugs, ganciclovir (Cytovene, 1988) and penciclovir (Denavir, 1996), followed.

In fighting the COVID-19 pandemic, Dr. Anthony Fauci compared remdesivir's efficacy against COVID-19, although not overwhelming, to the discovery of AZT for HIV for a good reason. After AZT, many "me-too" and "me-better" nucleoside reverse transcriptase inhibitors (NRTIs) emerged on the market. They include d4t (Zerit); ddI (Videx); lamivudine (3TC, Epivir), discovered by Dennis Liotta at Emory; and abacavir (Ziagen), discovered by GSK and Bob Vince at Minnesota. Meanwhile, many non-nucleoside reverse transcriptase inhibitors (NNRTIs) were also discovered: nevirapine (Viramune), efavirenz (Sustiva), delavirdine (Rescriptor), etravirine (Intelence), and rilpivirine (Edurant). Concurrently, many HIV protease inhibitors, such as saquinavir (Invirase), indinavir (Crixivan), ritonavir (Norvir), tipranavir (Aptivus), and darunavir (Prezista), have been discovered as well. In 1996, highly active antiretroviral therapy, also known as a cocktail of HIV drugs, transformed AIDS from a death sentence to a chronic disease that can be managed with drugs (close to a cure).

Closer to home, HCV drugs built the foundation for the discovery of remdesivir. Hepatitis C virus (HCV) drugs represent a major triumph of modern medicine. Different from hepatitis A virus (HAV) and hepatitis B virus (HBV), some patients affected by HCV have no symptoms at all, which explains why it was not until 1989 when HCV was cloned, even though about 200 million people worldwide are infected.

Unlike HSV and HIV, which are both DNA viruses, HCV is a single-strand RNA virus. Proteins encoded in the HCV genome include several non-structure (NS) proteins. One of them is the NS5B protein, an *RNA-dependent RNA polymerase* (RdRp; incidentally, remdesivir inhibits COVID-19's RdRp as well), which is the weakest link of the virus genome and the most vulnerable point for viral replication within the host. HCV NS5B, as well as other non-structural proteins, such as NS3/4A and NS5A, are responsible for the replication of the viral genome that represents important and tractable targets for drug design and development.

One of the shining stars of HCV NS5B inhibitors was sofosbuvir (Sovaldi), discovered by Pharmasset[7] and sold by Gilead. It captured every American's attention when it was sold at $1,000 per pill! The cost was justified by Gilead because sofosbuvir offered a cure when there was none, and the cost was still significantly less than any other alternatives at the time.

Sofosbuvir is a prodrug of a unique nucleoside, PSI-6130, which is fascinating in its own right because it has a rare fluorine-containing tertiary carbon at the $2'$-position of its ribose ring. This is a testimony of how important the fluorine atom has become in modern drug discovery. While no fluorine-containing drugs existed before fludrocortisone was approved in 1955, more than twenty percent of all drugs now contain one or more fluorine atoms. For the record, as of writing, the trophy for the drug with the most fluorine atoms (seven) goes to aprepitant (Emend), a drug marketed by Merck since 2003 to prevent nausea and vomiting brought about by cancer chemotherapy. It is a substance P antagonist and a neurokinin 1 (NK_1) inhibitor.

The major aspect that made sofosbuvir significantly superior than its parent drug, PSI-6130, is that the former is a prodrug of the latter. Browse any medicinal chemistry book and you will find that the prodrug strategy can turn a terrible drug into a decent one. The litany of prodrug benefits includes:

- overcoming formulation and administration problems,
- overcoming absorption barriers,
- overcoming distribution problems,
- overcoming metabolism and excretion problems, and
- overcoming toxicity problems.

As a consequence, prodrugs currently constitute five percent of known drugs and a larger percentage of new drugs.

Sofosbuvir is not just any prodrug—it is a state-of-the-art *phosphoramidate* prodrug. Those *Pro*drugs of nucleo*Tides* are also known as ProTides. The technology was developed in 1990 by Dr. Chris McGuigan at Cardiff University in Wales. It is probably the most successful prodrug strategy applied in the antiviral field.

To convert PSI-6130 to the active PSI-6130-TP, the phosphorylation has to take place first to make the 5′-monophosphate of PSI-6130. However, as it so often happens, the virus either does not induce a specific kinase or has developed resistance to the compound through mutations in this enzyme while human cells fail to secure phosphorylation. To overcome this issue, it is better to install a phosphate onto the nucleoside (which is now a nucleotide because it has a phosphate). Regrettably, phosphates are negatively charged, and those nucleotides all have poor cell penetration at a physiological pH. Among many tactics to overcome the polarity issue, the most successful of all is the ProTide, where the nucleotide is masked with an amino acid ester pro-moiety linked via a P–N bond. A frequently used amino acid, as in the case for sofosbuvir, is L-alanine. Such a ProTide can enter the cell via facilitated passive diffusion through the cell membrane. Once inside the cell, the monophosphate nucleoside is released and does what it supposed to do: it serves as a viral RNA-replication terminator. Sofosbuvir is such a successful pro-drug that it can be taken orally.

Meanwhile, Gilead came up with their own HCV NS5B drugs, one of which was nucleoside GS-44154, which has two points of differentiation in comparison to Pharmasset's PSI-6130. One difference is that the former has a tertiary carbon with the key nitrile group at the 1′-position on the ribose ring. More significantly, in place of the latter's natural N-nucleoside base, the former has an unnatural C-nucleoside base, also at the 1′-position on the ribose ring. Thus, GS-44154 is a 1′-cyano-substituted C-nucleoside ribose analog. In theory and in practice, C-nucleosides are more resistant to metabolism in human body. In 2012, while evaluating GS-44154 in cell-based assays against a panel of RNA viruses, GS-44154 was found to display a broad spectrum of activity against HCV, Dengue virus-2, influenza A, SARS-CoV, etc.[8]

Most nucleosides, including GS-44154, are poorly cell-permeable so that they can have a low hit rate in cell-based antiviral screens. Prodrugs are often deployed to enhance their cell permeability. For nucleoside GS-44154, after installation of a phosphoramidate, the prodrug *du jour*, Gilead arrived at their own HCV NS5B inhibitor, GS-6620. It was the first C-nucleoside HCV polymerase inhibitor with demonstrated antiviral response in HCV-infected patients.[9] With the overwhelming success of sofosbuvir (Sovaldi), although it was from the NIH (not invented in the United States), there was no need for another "me-too" drug for Gilead. The home-grown PSI-6130 languished on the self of Gilead's compound management.

Then came along the Ebola virus (EBOV), a member of the *Filoviridae* family. EBOV is a single-stranded, negative-sense, non-segmented RNA virus. During 2010s, more than twenty-eight thousand cases of Ebola virus disease were recorded in West Africa. To combat the pandemic, Gilead collaborated with the Center for Disease Control (CDC) and the United States Army Medical Research Institute of Infectious Diseases (USAMRIID). Together, they screened an assembly of approximately one thousand compounds, largely nucleosides and nucleotides, from Gilead's collection. They heavily focused on ribose analogs that could target RNA viruses because this would encompass many emerging viral infections ranging from respiratory pathogens belonging to the *Coronaviridae* family, such as severe acute respiratory syndrome (SARS) and Middle East respiratory syndrome (MERS), to mosquito-borne viruses of the *Filoviridae* family, such as Dengue and Zika. What emerged from the work was nucleoside GS-44154 and the corresponding phosphoramidate prodrug of its monophosphate. The prodrug was the S_p isomer GS-5734, which would go on to become remdesivir and Veklury in due course.[10] As any chemist would know, even a minute chemical structural change may result in tremendous differences in physiochemical and pharmacological properties. Despite striking structural similarities, remdesivir must be given via injection, whereas sofosbuvir is orally bioavailable.

In a 2012 study, remdesivir showed antiviral activity against SARS strain Toronto-2 without cytotoxicity toward the host cells. In 2016, therapeutic efficacy of remdesivir was demonstrated against Ebola virus in rhesus monkeys, a validated animal model.[11] Because the drug worked on nonhuman primates, it was deemed safe and efficacious enough to bring up to humans for clinical trials. At the time, four drugs were in clinical trials for treating the Ebola virus disease: ZMapp by Mapp Biopharmaceuticals, mAb114 by Vaccine Research Center at NIH, REGN-EB3 by Regeneron, and remdesivir by Gilead. Remdesivir is a small-molecule drug, while the other three drugs, ZMapp, mAb114, and REGN-EB3, are all monoclonal antibody (mAb) products. Monoclonal antibodies are derived from immune system molecules that bind to a specific substance, such as an invading virus. Pretty soon, ZMapp and remdesivir were dropped from the remainder of the trials because these two drugs were much less effective at preventing death than mAb114 and REGN-EB3. Eventually, it seemed that the Ebola virus pandemic was under control after ravaging West Africa for many years.

Then came along COVID-19 in Wuhan, China at the end of 2019. In terms of damages, COVID-19 takes the crown of all coronaviruses. Like the Ebola virus, SARS-CoV-2 is an RNA virus. It did not take long for Gilead scientists to propose that remdesivir could be a potential treatment of COVID-19 very early on. First, remdesivir showed inhibitory effects on pathogenic animal (rhesus macaques) and human coronaviruses, including SARS-CoV-2, the causative viral pathogen of COVID-19, *in vitro* by inhibiting its RNA-dependent RNA polymerase (RdRp). It also inhibits SARS and MERS coronaviruses.

Ironically, remdesivir sounds like "people's hope" in Chinese phonetically. During the dark days of the initial pandemic, remdesivir was people's only hope.

Although the outcome of clinical trials for remdesivir in China was ambiguous, Dr. Fauci's team at National Institute of Allergy and Infectious Diseases (NIAID) found statistically significant benefits of remdesivir for treating COVID-19 patients using the gold standard of clinical trials: randomized and double-blinded clinical trials. The rest, like they say, is history.[12]

Remdesivir, a bioisostere of natural N-nucleosides such as adenosine and guanosine, is a C-nucleoside antiviral drug, which works as an antimetabolite and blocks the synthesis of viral RNA. There are four RNA bases: adenine, guanine, cytosine, and uracil. Another base, thymine, is only seen in DNA. Remdesivir's pyrrolotriazine fragment bears a striking resemblance to adenine and guanine. Therefore, pyrrolotriazine can serve as a bioisostere of these two bases required for the synthesis of viral RNAs.

The discovery of remdesivir took a long and winding route. Initially discovered as a treatment of Ebola virus (EBOV),[13] it evolved from GS-6620, the first C-nucleoside HCV polymerase inhibitor with demonstrated antiviral response in HCV-infected patients. C-nucleosides have the potential for improved metabolism and pharmacokinetic properties over their natural N-nucleoside counterparts because of the presence of a strong carbon–carbon glycosidic bond and a non-natural heterocyclic base.

In terms of mechanism of action (MoA), remdesivir belongs to a class of nucleoside antiviral drugs called antimetabolites. By definition, antimetabolites are drugs that interfere with normal cellular function, particularly the synthesis of RNA that is required for replication. In another word, remdesivir "pretends" to be a nucleoside building block and participates in the viral RNA synthesis. But because it is not a *bona fide* RNA building block, RNA replication is terminated, and the virus is stopped from replicating.

But the reality is not that simple. Remdesivir is a prodrug that is only active *in vivo* in the form of its metabolite: triphosphate nucleoside, which has a low solubility and low bioavailability and could even be toxic. To overcome these shortcomings, the prodrug tactic was used, but with a twist. To achieve an optimal pharmacokinetic and pharmacodynamic outcome, several prodrug tactics are installed onto triphosphate nucleoside. Therefore, remdesivir may be considered as a prodrug of a prodrug of a prodrug of triphosphate nucleoside, the active antimetabolite.

Once remdesivir is injected via IV into the human body, the key enzymes initially involved in the metabolism of remdesivir are human cathepsin A 1 (CatA1) and carboxylesterase 1 (CES1), which are responsible for the hydrolysis of the carboxyl ester between the alaninyl moiety and 2-ethyl-2-butanol. This stereospecific reaction gives rise to the corresponding carboxylic acid. A non-enzymatic intramolecular nucleophilic attack then results in the formation of an alaninyl phosphate intermediate, which undergoes a rapid chemical reaction to hydrolyze the cyclic phosphate to a linear phosphate as carboxylic acid, along with a phenol. The next step is speculated to involve the histidine triad nucleotide-binding protein 1 (Hint 1) enzyme in which the alaninyl phosphate intermediate is deaminated to form a monophosphate nucleotide. The final two steps involve consecutive phosphorylation reactions mediated by cellular kinases, uridine monophosphate–cytidine monophosphate kinase (UMP–CMPK), and nucleoside diphosphate kinase (NDPK), producing the diphosphate nucleotide and subsequently the active triphosphate nucleotide, respectively.[14] Triphosphate nucleotide is the nucleoside antimetabolite and is incorporated to the viral DNA to stop its replication.

Now with the FDA's approval, Gilead is ramping up remdesivir's production, made easier by their process chemists' heroic efforts in optimizing the manufacturing route. Doctors can start prescribing remdesivir for emergency use in COVID-19 patients now and more and more data will accumulate on remdesivir's safety and efficacy profile in a large population of patients.[15]

Remdesivir is not a silver bullet, but with its proof-of-concept (PoC) in treating COVID-19 patients, we know this mechanism works. "Now this is not the end. It is not even the beginning of the end. But it is, perhaps, the end of the beginning," to quote Churchill's famous words. It is expected that many "me-too," hopefully "me-better" drugs and innovative drugs with novel MoAs will follow and will enable combination therapies with better safety and efficacy profiles.

5.3.2 Dexamethasone (Decadron)

On June 16, 2020, a team at Oxford University announced that their RECOVERY trials had revealed that a daily treatment of 6 mg of dexamethasone (Decadron) lowered the fatality rate of ventilated COVID-19 patients by up to one-third! This was a randomized, double-blinded clinical trial, the gold standard, thus lending much credibility to the results. It seems to work even better than Gilead's remdesivir (Veklury) for this group of seriously ill patients. Moreover, dexamethasone is cheap, widely available, and it has been used for more sixty years in the clinics.

Let us take a journey to look at how dexamethasone was discovered, how it has become almost a panacea for all ills in medical practice, how it works to benefit COVID-19 patients, and how it is made.

A group of chemists at Merck discovered dexamethasone, a steroid hormone, in 1958 as an anti-inflammatory steroid.[16] Merck marketed dexamethasone with a brand name Decadron in 1959. As a testimony to how competitive the field was during the golden age of steroid hormone drugs, Schering Corporation simultaneously synthesized the same compound as well. Considering the two companies merged in 2009, it matters not today with regard to priority.

Inflammation and immunity, like all other normal reactions of the body, are meant to preserve or restore health. Classic inflammatory diseases include rheumatoid arthritis and Crohn's disease (an inflammatory bowel disease). Rheumatoid arthritis is a chronic inflammatory disease characterized by pain, swelling, and subsequent destruction of joints. Philip S. Hench and Edward C. Kendall at the Mayo Clinic discovered cortisone and isolated it from bovine adrenal cortex (a small organ attached to the top of the kidney; cortex means out-layer). In 1948, Hench gave cortisone to a desperately ill twenty-nine-year-old Mrs. Gardner with severe rheumatoid arthritis. After three days of injections, the long bedridden patient miraculously recovered. She even went downtown and had a three-hour shopping spree! That single event ushered *the cortisone era.*[17]

As an understanding of the functions of cortisone increased exponentially, it was soon realized that cortisone itself is a prodrug, which is reduced *in vivo* to cortisol (hydrocortisone). Cortisol, the actual active drug, is vital to the body's defense against inflammation. Because Addison's disease is due to adrenal cortex deficiency and is characterized by the failure of the adrenal glands and the inability to produce cortisol, cortisol became the drug of choice for treatment. President John F. Kennedy suffered from Addison's disease and was routinely given cortisol shots.

Because steroid hormones cortisone and cortisol were isolated from cortex, they were named as corticosteroids. There are two main groups of corticosteroids: glucocorticoids (good) and mineralocorticoids (bad). While cortisone and cortisol revolutionized medicines, they are far from perfect with many adverse effects, chief among them is mineralocorticoid activity (i.e., salt and water-retaining activity). They are also plagued by metabolic side effects, such as weight gain or increased blood glucose levels. Initially, scientists were afraid of tweaking their structures because "nature knows the best." In 1954, Josef Fried and Emily Sabo at Squibb debunked that notion by preparing semi-synthetic cortisone derivatives with different pharmacological profiles. For instance, their 9α-fluorohydrocortisone (fludrocortisone) was ten-fold more potent than cortisone in relieving rheumatoid arthritis (good) although their mineralocorticoid activity was increased by three-hundred-fold to eight-hundred-fold (bad, causing aldosterone-like salt and water retaining).[18] Because Fried and Sabo showed the world that modifying natural corticosteroids could lead to different, or even better, drugs, chemists in drug industry became *bona fide* medicinal chemists: they began to explore the structure–activity relationship (SAR). More excitingly, almost all the best organic chemists in the world joined the fray, many of them future Nobel laureates, such as R. B. Woodward, Robert Robinson, Derek Barton, and Carl Djerassi, among others.

In 1954, from the fermentation of cortisone in a mold broth *Corynebacterium simplex*, Schering Corporation isolated two novel steroids prednisone (C-11 ketone, brand name Meticorten) and the active metabolite prednisolone (C-11 beta hydroxyl) as a consequence of microbial dehydrogenation. These analogs exhibited enhanced glucocorticoid activity with reduced mineralocorticoid activity. The first dose of prednisolone was administered to an arthritic woman in August 1954. Prednisone was launched in March 1955 at a 5 mg dose for use in arthritis, with prednisolone being introduced shortly thereafter. Schering's initial $100K investment earned them $20 million in the first year of sales, a two-hundred-fold profit.

Both the prodrug prednisone and the active drug prednisolone are approximately five-fold more active than cortisol (hydrocortisone) with less tendency to cause salt retention.

At the time, the synthesis of steroids, large molecules with many chiral centers, was long and formidable. Medicinal chemists had no rational design guidelines to follow, and progress was dictated by stepwise improvements enabled by synthetic chemistry.

As showcased by fludrocortisone, halogen, especially fluorine, incorporation resulted in surprisingly active steroids, although fludrocortisone had increased undesirable mineralocorticoid activity. Lederle Laboratories isolated a corticosteroid from the urine sample of a boy taking cortisone. The compound had a C-16 hydroxyl group added to the cortisone molecule, no doubt from CYP450 metabolic oxidation. Merging the hydroxyl group and the fluorine substitution on fludrocortisone, Lederle arrived at a hybrid molecule, triamcinolone, in 1958. Triamcinolone was as potent as prednisolone but was almost free from mineralocorticoid activity. It is not perfect, though, as it has a tendency to cause nausea, dizziness, and other untoward effects.

In an attempt to hinder metabolism of the side chain corticosteroids (the ketone at the C-20 position was prone to reduction, thus losing activity), Merck chemists led by Lewis Hastings Sarett synthesized analogs with a methyl group at the adjacent C-16 position, such as dexamethasone. As is often the case in science, the 16-methyl did not really retard metabolism as they initially intended because CYP450 enzymes oxidatively metabolize methyl group with ease. However, unexpectedly, dexamethasone not only acted like a hydroxyl group to depress mineralocorticoid activity, but it also enhances the anti-inflammatory activity by six-fold in comparison to prednisone and was twenty-five times more potent than hydrocortisone.[19] Dexamethasone has a duration of action of thirty-six to fifty-four hours whereas that of hydrocortisone is eight to twelve hours.

Shortly after, Merck also synthesized betamethasone, its diastereomer with the methyl group at C-16 position with opposite stereochemistry. Betamethasone, a twin brother of dexamethasone in a true sense, behaves nearly identical to each other in terms of biological activities. Apparently, the stereochemistry of the C-16 methyl has little pharmacological consequences. More favorably, both dexamethasone and betamethasone have practically no salt-retaining activity at all. Neither of them exhibited the unwelcome side effects seen with triamcinolone.

Interestingly, although it is an anti-inflammatory medicine, dexamethasone can help mitigate the effect of altitude sickness and high-altitude edema. When taken prophylactically, it can help climbers ascend quickly. Not surprisingly, dexamethasone may be found in every climber's backpacker when they climb the Mount Everest.

Today dexamethasone is available as an IV, oral, nasal, ophthalmic, and topical cream formulation for treatment of diseases ranging from multiple myeloma to psoriasis. Although glucocorticoids are associated with osteoporosis, fluid retention, and hyperglycemia with chronic therapy, more acute therapy is remarkably efficacious and explains widespread use of steroidal therapy. Unfortunately, the initial enthusiasm for corticosteroids was dampened by their severe side effects following chronic administration. Notably osteoporosis, immune suppression, ulcerogenicity, adrenal suppression, and development of steroid dependence. Prednisone saves lives but takes an awful toll due to side effects, the most severe being the weakening of the bones within months. As a consequence, for the treatment of inflammatory diseases, corticosteroids have now been largely replaced by non-steroidal anti-inflammatory drugs (NSAIDS). Still, the fact remains that corticosteroids are currently the most effective drug in the treatment of acute and chronic inflammatory diseases.

In terms of its MoA, dexamethasone, like most corticosteroids, is very complicated, hitting multiple drug targets. But its key anti-inflammatory properties are the consequence of serving as a modulator (agonist) of a ligand-bound glucocorticoid receptor. Dexamethasone binds to glucocorticoid receptor on cell membrane. The dexamethasone–glucocorticoid receptor complex then enters the cell nucleus to interact with the nuclear DNA molecules that trigger the signaling pathway for an anti-inflammatory outcome. Unfortunately, dexamethasone's poly-pharmacology comes from its binding to several other targets, chief among them the mineralocorticoid receptor, which is responsible for mineralocorticoid side effects, including water and salt-retention.

Since its emergence on the market in 1959, dexamethasone has become one of the two most popular orally bioavailable corticosteroids. The other one is prednisone. Having been on the market for over sixty years, dexamethasone has been tried with a certain degree of success for many indications, making it seem like a panacea. For instance, it has been found to have efficacy for rheumatic problems, a number of skin diseases, severe allergies, asthma, chronic obstructive lung disease, croup, brain swelling, and, along with antibiotics, in tuberculosis. Doctors are very familiar with dexamethasone's pharmacological properties. Therefore, they have been using this drug on severe COVID-19 patients "off label" to relieve their pneumonia symptoms. The RECOVERY clinical trial simply confirmed what doctors inferred early on at the peak of the pandemic, and now they can start using dexamethasone with much more confidence.

In summary, dexamethasone has had a colorful "career" as a drug. Initially discovered as an anti-inflammatory drug, it has emerged as one of the two orally bioavailable corticosteroid medicines. It has been used to combat nausea associated with chemotherapy, as an anesthetic, and an anti-allergic. The drug has also been found useful in the treatment of asthma and chronic obstructive pulmonary disease (COPD). Now, as an effective medicine to reduce the fatality rate of ventilated COVID-19 patients, dexamethasone continues to save human lives.

Fig. 5.4 Coronavirus © Uruguay Post

6

Closing Remarks

The more you know, the more you know about what you don't know.

—*Aristotle*

In this book, we have reviewed several classes of viruses that are common pathogens, including:

- HIV, RNA retrovirus;
- HAV, RNA virus;
- HBV, DNA retrovirus;
- HCV, RNA virus;
- Influenza virus, RNA virus; and
- Coronavirus, RNA virus.

With the exception of the hepatitis B virus, all other frequently encountered viruses are RNA viruses. RNA viruses mutate frequently and evolve in vast animal reservoirs. We have reviewed these viruses, their vaccines, and their therapeutic drugs. I hope history will shed light on, and help us to understand, today's COVID-19 pandemic.

Normally, influenza is our perennial virus concern. For influenza, human infection with the highly pathogenic avian-origin H5N1 influenza A virus, commonly referred to as "avian flu," has been significantly more severe and associated with a very high mortality rate. Influenza pandemics should have served as lessons

Conquest of Invisible Enemies. Jie Jack Li, Oxford University Press. © Oxford University Press 2022.
DOI: 10.1093/oso/9780197609859.003.0006

learned for us, especially the lessons from the 1918 pandemic. Jeffery Taubenberger stated in 2018: "The possibility that more than two million people could suddenly need intensive care with ventilatory support is a frightening reminder of the challenge of an influenza pandemic."[1]

Little did Taubenberger know, his fear would come true in mere two years, although it did not come as an influenza pandemic, but instead as a coronavirus pandemic. The need of intensive care with ventilatory support has, so far, exceeded three million people. Thankfully, vaccines are saving the day, with over 168 million American adults having received the vaccine by June 2021. The pandemic is largely under control in the United States.

However, the last two years have not been normal with coronavirus ravaging around the globe. Even today (June 1, 2021), we are living in a precarious time. China has had COVID-19 under control largely through aggressive public health measure. Although vaccination has started in China, it will take a long time to achieve herd immunity for a country of 1.4 billion. The United States, on the other hand, has suffered a stunning 33.3 million cases and over six hundred thousand fatalities thus far, despite being the richest country with the best medical resource in the world. Thankfully, three vaccines, developed by Pfizer/BioNTech, Moderna, and Johnson & Johnson/Janssen, have saved millions of American lives while other vaccines aim to serve other international populations. On June 1, 2021, the United States saw only about 22,542 cases and 520 deaths. Regrettably, other parts of the world are still ravaged by the coronavirus, especially India and Brazil.

With vaccination and public health measures, I hope it will not be long until we completely vanquish coronavirus, the invisible enemy.

Science will win!

Chemical Structures of the Drugs

Chapter 2. HIV/AIDS Drugs: Transforming a Death Sentence into a Chronic Disease

Thymidine synthetic thymidine nucleoside analogs

thymidine

idoxuridine
(IdU. Herplex). 1962

trifluorothymidine
(TFT. Viroptic, GSK) 1980

ethyldeoxyuridine
(EdU): 1985

bromovinyldeoxyuridine
(brivudine, Zostex), 2001

telbivudine
(Tyzeka, Novartis) 2006

HIV Nucleoside Reverse Transcriptase Inhibitors

zidovudine (AZT, azidothymidine,
Retrovir, Burroughs Wellcome). 1987

didanosine
(ddI, Videx, BMS), 1991

stavudine
(d4T, Zerit,
BMS), 1994

lamivudine
(3TC, Epivir,
GSK), 1995

abacavir
(Ziagen,
GSK), 1999

tenofovir disoproxil (Viread)
Gilead, 2001

emtricitabine (FTC, Emtriva)
Gilead, 2003

Non-Nucleoside Reverse Transcriptase Inhibitors

nevirapine (Viramune)
Boehringer Ingelheim
Launched: 1996

delavirdine (Rescriptor)
Upjohn/Pfizer
Launched: 1997

efavirenz (Sustiva)
Bristol-Myers
Squibb/Merck
Launched: 1998

etravirine (Intelence)
Tibotec/J&J, 2008

rilpivirine (Edurant)
Tibotec/J&J, 2011

doravirine (Pifeltro)
Schering Plough/Merck, 2018

HIV Protease Inhibitors

saquinavir (Invirase)
Roche, 1995

indinavir (Crixivan)
Merck 1996

ritonavir (Norvir)
Abbott, 1996

nelfinavir (Viracept)
Agouron/Eli Lilly, 1997

R = H amprenavir (Agenerase, GSK), 1999
R = PO(OH)$_2$ fos-amprenavir (Lexiva, GSK/Vertex), 2005

lopinavir with ritonivir (Kaletra)
Abbott, 2000

atazanavir (Reyataz)
Ciba Geigy, 2003

tipranavir (Aptivus)
Boehringer Ingelheim, 2005

darunavir (Prezista)
Tibotec, 2006

HIV Integrase Inhibitors

raltegravir (Isentress)
Merck, 2007

dolutegravir (Trivicay)
GSK, 2013

elvitegravir (Vitekta)
Japan Tabacco/Gilead, 2014

bictegravir (Biktarvy with
emtricitabine/tenofovir alafenamide)
Gilead, 2018

cabotegravir (HIV integrase
inhibitor, an ingredient of
Carbenuva, with rilpivirine)
ViiV/GSK, 2020

HIV Entry Inhibitor

maraviroc (Selzentry)
Pfizer, 2007
CCR5 Antagonist

Chapter 3. Hepatitis Viruses

Nucleoside HBV drugs

Lamivudine
(3TC, Epivir)
GSK, 1995

adefovir dipivoxil
(Hepsera)
Gilead, 2002

entecavir (Baraclude)
BMS, 2005

telbivudine (Tyzeka)
Novartis, 2006

tenofovir disoproxil (Viread)
Gilead, 2008

alafenamide (Vemlidy)
Gilead, 2016

HCV NS3/4A Serine Protease Inhibitors

boceprevir (Victrelis)
Schering-Plough/Merck, 5-13-2011

telaprevir (Incivek)
Vertex, 5-23-2011

ciluprevir (BILN2061)
Boehringer Ingelheim
Discontinued in 2004

faldaprevir (BILN201335)
Boehringer Ingelheim
Discontinued in 2014

narlaprevir (Arlansa)
R-Pharm/Merck, 2016
Russia Only

simeprevir (Olysio)
Tibotec/Medivir/Janssen, 2013

vaniprevir (Vanihep)
Merch, 2014, Japan

danoprevir (Ganovo)
Roche, 2013 (China)

paritaprevir
Abbvie, 2014

asunaprevir (Sunvepra)
BMS, 2014

grazoprevir (Zepatier, combo)
Merch, 2016

glecaprevir (Mavyret, combo)
Abbvie, 2017

voxilaprevir (Vosevi, combo)
Gilead, 2017

HCV NS5A Protein Inhibitors

daclatasvir (Daklinza)
BMS, 2015

ledipasvir (Harvoni, combo)
Gilead, 2014

ombitasvir (Viekira Pak, 2014)
Abbvie, 2014

elbasvir (Zeptier with Grazoprevir)
Merch, 2015

velpatasvir
(Vosevi, combo)
Geilad, 2016

pibrentasvir (Mavyret, combo)
Abbvie, 2017

HCV NS5B Polymerase Inhibitors

sofosbuvir (Sovaldi)
Gilead, 2013

dasabuvir (Exviera, combo)
Abbvie, 2014

beclabuvir (Ximency, combo)
BMS, 2016

HCV NS5A Protein Inhibitors

H₃CO₂CHN ... daclatasvir (Daklinza)
BMS, 2015

Chapter 4. Influenza: A Perennial Killer

amantadine (Symmetrel)
Smith Kline and French, 1968
M2 ion channel inhibitor

rimantadine (Flumadine)
Forest Pharmaceuticals, 1994
M2 ion channel inhibitor

DANA

zanamivir (Relenza)
GSK, 1999
neuraminidase inhibitor

oseltamivir (Tamiflu)
Gilead/Roche, 1999
neuraminidase inhibitor

peramivir (Rapivab)
BioCryst, 2015
neuraminidase inhibitor

laninamivir (Inavir)
Sankyo, 2013
neuraminidase inhibitor

laninamivir octanoate

Baloxavir acid (BXA)

Baloxavir marboxil (BXM)
Shionogi/Genentech/Roche, 2018
RNA polymerase inhibitor

Chapter 5. Coronaviruses

remdesivir (Veklury)
Gilead, 2020
C-Nucleoside antiviral

dexamethasone (Decardon)
Merck, 1959
glucocorticoid receptor modulator

Notes

Chapter 1. Viruses Shaping History and the Discovery of Viruses

1. Mike Dash, *Tulipomania: The Story of the World's Most Coveted Flower & the Extraordinary Passions It Aroused* (New York: Broadway Books, 2001).
2. E. L. Dekker, A. F. Derks, C. J. Asjes, et al., "Characterization of Potyviruses from Tulip and Lily Which Cause Flower-Breaking," *The Journal of General Virology* 74, no. 5 (1993): 881–888.
3. Mike Dash, *Tulipomania: The Story of the World's Most Coveted Flower & the Extraordinary Passions It Aroused* (New York: Broadway Books, 2001).
4. Judith A. Lesnaw and Said A. Ghabrial, "Tulip Breaking: Past, Present, and Future," *Plant Disease* 84, no. 10 (2000): 1052–1060.
5. Mike Dash, *Tulipomania: The Story of the World's Most Coveted Flower & the Extraordinary Passions It Aroused* (New York: Broadway Books, 2001).
6. Thomas Hugh, *Conquest, Montezuma, Cortés, and the Fall of Old Mexico* (New York: Simon & Schuster, 1993).
7. Catherine Théves, Eric Crubézy, and Philippe Biagini, "History of Smallpox and Its Spread in Human Populations," *Microbiology Spectrum* 4, no. 4 (2016): 1–10.
8. Kristine B. Patterson and Thomas Runge, "Smallpox and the Native American," *The American Journal of the Medical Sciences* 323, no. 4 (2002): 216–222.
9. Kim MacQuarrie, *The Last Days of The Incas* (New York: Simon and Schuster Paperbacks, 2007).
10. Kristine B. Patterson and Thomas Runge, "Smallpox and the Native American," *The American Journal of the Medical Sciences* 323, no. 4 (2002): 216–222.
11. Kim MacQuarrie, *The Last Days of The Incas* (New York: Simon and Schuster Paperbacks, 2007).
12. James C. Riley, "Smallpox and the American Indians Revisited," *Journal of the History of Medicine and Allied Sciences* 64, no. 4 (2010): 445–477.
13. Neil Metcalf, "A Short History of Biological Warfare," *Medicine, Conflict and Survival* 18, no. 3 (2002): 271–282.

14. Chia-Feng Chang, "Disease and Its Impact on Politics, Diplomacy, and the Military: The Case of Smallpox and the Manchus (1613–1795)," *Journal of the History of Medicine and Allied Sciences* 57, no. 2 (2002): 177–197.

15. Philip H. Clendenning, "Dr. Thomas Dimsdale and Smallpox Inoculation in Russia," *Journal of the History of Medicine and Allied Sciences* 28, no. 2 (1973): 109–125.

16. Charles A. Cerami, *Jefferson's Great Gamble, The Remarkable Story of Jefferson, Napoleon and the Men behind the Louisiana Purchase* (Naperville, IL: Sourcebooks, Inc., 2003).

17. René Dubos, *Pasteur and Modern Science, Scientific Revolutionaries: A Biographical Series* (Madison, WI: Science Tech Publishers, 1960); John Waller, *The Discovery of the Germ: Twenty Years That Transformed the Way We Thank About Disease* (New York: Columbia University Press, 2002).

18. Gerald L. Geison, *The Private Science of Louis Pasteur* (Princeton, NJ: Princeton University Press, 1995.

19. Alice Lustig and Arnold Levine, "One Hundred Years of Virology," *Journal of Virology* 66, no. 8 (1992): 4629–4631; Andrew W. Artenstein, "The Discovery of Virus: Advancing Science and Medicine by Challenging Dogma," *International Journal of Infectious Diseases* 16, no. 7 (2012): e470–e473.

20. Theodor O. Diener, "Discovery of Viroids—A Personal Perspective," *Nature Review, Microbiology* 1, no. 1 (2003): 75–80.

21. Robin A. Weiss and Peter K. Vogt, "100 Years of Rous Sarcoma Virus," *Journal of Experimental Medicines* 208, no. 12 (2011): 2351–2355; Eva Becsei-Kilborn, "Scientific Discovery and Scientific Reputation: The Reception of Peyton Rous' Discovery of the Chicken Sarcoma Virus," *Journal of the History of Biology* 43 (2010): 111–157.

22. Dorothy H. Crawford, Alan Rickinson, and Ingólfur Johannessen, *Cancer Virus, The Story of Epstein-Barr Virus* (Oxford, UK: Oxford University Press, 2014).

23. L. S. Young, L. F. Yap, and P. G. Murray, "Epstein-Barr Virus: More Than 50 Years Old and Still Providing Surprises," *Nature Reviews, Cancer 16* (2016): 789–802.

24. Anthony Epstein, "Why and How Epstein-Barr Virus Was Discovered 50 Years Ago," *Current Topics in Microbiology and Immunology* 390 (2015): 3–15.

25. M. A. Epstein, B. G. Achong, and Y. M. Barr, "Virus Particles in Cultured Lymphoblasts from Burkitt's Lymphoma," *Lancet* 1 (1964): 702–703.

26. R. T. Javier and J. S. Butel, "The History of Tumor Virology," *Cancer Research* 68, no. 19 (2008): 7693–7706.

27. J. M. Coffin and H. Fan, "The Discovery of Reverse Transcriptase," *Annu. Rev. Virol. 3* (2016): 29–751.

28. Mitsuaki Yoshida, "Discovery of HTLV, The First Human Retrovirus, Its Unique Regulatory Mechanisms, and Insight into Pathogenesis," *Oncogen*, 24 (2005): 5931–5937.

29. Robert C. Gallo, "A Historical Personal Perspective on Human Retroviruses and Their Infection on T Cells," *Transfusion 55*, (2015): 1–9.

Chapter 2. HIV/AIDS Drugs: Transforming a Death Sentence into a Chronic Disease

1. UN/UNAIDS, "Statistics," https://www.unaids.org/en/resources/fact-sheet, accessed on September 9, 2020.

2. Robert C. Gallo, *Virus Hunting, AIDS, Cancer, and the Human Retrovirus: A Story of Scientific Discovery* (New York: Basic Books, 1991).

3. Luc Montagnier, *Virus, The Codiscoverer of HIV Tracks Its Rampage and Charts the Future* (New York: W. W. Norton & Company, 2000);
Luc Montagnier, "25 Years after HIV Discovery: Prospects for Cure and Vaccine (Nobel Lecture)," *Angew. Chem. Int. Ed.* 48 (2009): 5815–5826;
Françoise Barré-Sinoussi, "HIV: A Discovery Opening the Road to Novel Scientific Knowledge and Global Health Improvement (Nobel Lecture)," *Angew. Chem. Int. Ed.* 48 (2009): 5809–5814.

4. Daniel S. Greenberg, "Resounding Echoes of Gallo Case," *The Lancet* 345 (1995): 639; Linda Marsa, *Prescription for Profits, How the Pharmaceutical Industry Bankrolled the Unholy Marriage Between Science and Business* (New York: Scribner, 1997).

5. John Crewdson, "The Great AIDS Quest," *Chicago Tribute*, November 19, 1999.

6. Robert C. Gallo, *Virus Hunting, AIDS, Cancer, and the Human Retrovirus: A Story of Scientific Discovery* (New York: Basic Books, 1991).

7. John Crewdson, *Science Fictions: A Scientific Mystery, a Massive Cover-up and the Dark Legacy of Robert Gallo* (New York: Back Bay Books, 2003).

8. Robert C. Gallo, *Virus Hunting, AIDS, Cancer, and the Human Retrovirus: A Story of Scientific Discovery* (New York: Basic Books, 1991).

9. Randy Shilts, *And The Band Played On: Politics, People, and the AIDS Epidemic (1980–1985)* (New York: St. Martin's Griffin, 1987).

10. István Hargittai, *The Road to Stockholm, Nobel Prizes, Science, and Scientists* (Oxford: Oxford University Press, 2002).

11. Giovanni Abbadessa, et al., "Unsung Hero Robert C. Gallo," *Science* 323, no. 5911 (2009): 206–207.

12. Aners Vahlne, "A Historical Reflection on the Discovery of Human Retrovirus," *Retrovirology* 6 (2009): 40.

13. Michael Bliss, *The Discovery of Insulin* (2nd ed.) (Toronto: University of Toronto Press: 1982); Michael Bliss, *Banting: A Biography* (Toronto: University of Toronto Press, 1992).

14. P. D. Stolley and T. Lasky, "Johannes Fibiger and His Nobel Prize for the Hypothesis That A Worm Causes Stomach Cancer," *Annals of Internal Medicine* 116 (1992): 765–769; I. M. Modlin, M. Kidd, and T. Hinoue, "Of Fibiger and Fables," *Journal of Clinical Gastroenterology 33*, no. 3 (2001): 177–179; Carl-Magnus Stolt, George Klein, and T. R. Jansson, "An Analysis of a Wrong Nobel Prize—Johannes Fibiger, 1926: A Study in the Nobel Archives," *Advances in Cancer Research 92* (2004): 1–12.

15. D. Stehelin, H. E. Varmus, J. M. Bishop, et al., "DNA Related to the Transforming Gene(s) of Avian Sarcoma Viruses is Present in Normal Avian DNA," *Nature* 260 (1976): 170–173.

16. J. Marx, "Stehlin Persists in Nobel Protest," *Science* 246, no. 4934 (1989): 1121.

17. Robert A. Weinberg, *Racing to the Beginning of the Road, The Search for the Origin of Cancer* (New York: W. H. Freeman and Company, 1998).

18. W. H. Prusoff and D. C. Ward, "Nucleoside Analogs with Antiviral Activity," *Biochemical Pharmacology* 25 (1976): 1233–1239; Raymond F. Schinazi, "William H. Prusoff Dies at 90—A Renaissance Man that Revolutionized the Treatment of Herpesviruses and HIV," *Antiviral Chemistry and Chemotherapy 21* (2011): 219–220.

19. Raymond F. Schinazi, "William H. Prusoff Dies at 90—A Renaissance Man that Revolutionized the Treatment of Herpesviruses and HIV," *Antiviral Chemistry & Chemotherapy* 21 (2011): 219–220.

20. V. K. Sharma, R. K. Sharma, P. K. Singh, et al., "An Engrossing History of Azidothymidine," *Immunol. Endocr. Metab. Agents Med. Chem.* 15, no. 2 (2015): 168–175; Samuel Broder, "The Development of Antiretroviral Therapy and its Impact on the HIV-1/AIDS Pandemic," *Antiviral Research* 85 (2010): 1–18; R. Yarchoan and Samuel Broder, "Development of Antiretroviral Therapy for the Acquired Immunodeficiency Syndrome and Related Disorders. A Progress Report," *N. Engl. J. Med. 316*, no. 9 (1987): 557–564.

21. J. C. Martin, M. J. M. Hitchcock, E. De Clercq, et al., "Early Nucleoside Reverse Transcriptase Inhibitors for the Treatment of HIV: A Brief History of Stavudine (D4T) and Its Comparison with Other Dideoxynucleosides," *Antiviral Research* 85 (2010): 34–38; Erik De Clercq, "A 40-year Journey in Search of Selective Antiviral Chemotherapy," *Annual Review of Pharmacology and Toxicology 51* (2011): 1–24.

22. Dennis C. Liotta and George R. Painter, "Discovery and Development of the Anti-Human Immunodeficiency Virus Drug, Emtricitabine (Emtriva, FTC)," *Accounts of Chemical Research* 49, no. 10 (2016): 2091–2098; D. Anason, "Emory and Glaxo Escalate was Over Hottest AIDS Drug," *Atlanta Business Chronicle*, August 5, 1996.

23. Karl Grozinger, John Proudfoot, and Karl Hargrave, "Discovery and Development of Nevirapine," in *Drug Discovery and Development*, edited by Mukund S. Chorghade, 353–363 (Hoboken, NJ: John Wiley and

Sons, 2006); Karl D. Hargrave, John R. Proudfoot, Karl G. Grozinger, et al., "Novel Non-Nucleoside Inhibitors of HIV-1 Reverse Transcriptase. 1. Tricyclic pyridobenzo- and dipyridodiazepinones," *Journal of Medicinal Chemistry 34*, no. 7 (1991): 2231–2241.

24. Terry A. Lyle, "Discovery of Indinavir and Efavirenz, New Therapeutic Agents for AIDS," *Chimia* 53, no. 6 (1999): 295–296.

25. Koen Andries, Ann Debunne, Thomas N. Kakuda, et al., "Etravirine: From TMC125 to Intelence: A Treatment Paradigm Shift for HIV-infected Patients," in *Antiviral Drugs*, edited by Wieslaw M. Kazmierski, 71–84 (Hoboken, NJ: John Wiley and Sons, 2011).

26. Ian B. Duncan, and Sally Redshaw, "Discovery and Early Development of Saquinavir," *Infectious Diseases Therapy* 25 (2002): 27–47.

27. Dale J. Kempf, "Discovery and Early Development of Ritonavir and ABT-378," *Infectious Diseases Therapy* 25 (2002): 49–64.

28. S. R. Chemburkar, J. Bauer, K. Deming, et al., "Dealing with the Impact of Ritonavir Polymorph on the Last Stage of Bulk Drug Process Development," *Organic Process Research & Development* 4 (2000): 413–417.

29. L. Xu, H. Liu, B. P. Murray, et al., *ACS Med. Chem. Lett. 1* (2010): 209–213.

30. L. Xu, H. Liu, B. P. Murray, et al., *ACS Med. Chem. Lett.* 1 (2010): 209–213.

31. L. Xu, H. Liu, B. P. Murray, et al., *ACS Med. Chem. Lett.* 1 (2010): 209–213.

32. Bruce D. Dorsey and Joseph P. Vacca, "Discovery and Early Development of Indinavir," *Infectious Diseases Therapy* 25 (2002): 65–83.

33. Karen R. Romines, "Discovery and Development of Tipranavir," in *Antiviral Drugs*, edited by Wieslaw M. Kazmierski, 47–57 (Hoboken, NJ: Wiley, 2011).

34. Arun K. Ghosh, Perali Ramu Sridhar, Nagaswamy Kumaragurubaran, et al., "Bis-tetrahydrofuran: A Privileged Ligand for Darunavir and A New Generation of HIV Protease Inhibitor that Combat Drug Resistance," *ChemMedChem* 1, no. 9 (2006): 939–950.

35. Arun K. Ghosh, Perali Ramu Sridhar, Nagaswamy Kumaragurubaran, et al., "Bis-tetrahydrofuran: A Privileged Ligand for Darunavir and A New Generation of HIV Protease Inhibitor that Combat Drug Resistance," *ChemMedChem* 1, no. 9 (2006): 939–950.

36. Michael Rowley, "The Discovery of Raltegravir, an Integrase Inhibitor for the Treatment of HIV Infection," *Prog. Med. Chem.* 46 (2008): 1–28.

37. Hisashi Shinkai, Masanori Sato, and Yuji Matsuzaki, "Elvitegravir: A Novel Monoketo Acid HIV-1 Integrase Strand Transfer Inhibitor," in *Antiviral Drugs*, edited by Wieslaw M. Kazmierski, 197–205 (Hoboken, NJ: Wiley, 2011).

38. David Price, "Maraviroc (Selzentry): The First-in-Class CCR5 Antagonist for the Treatment of HIV," in *Modern Drug Synthesis*, edited by Jie Jack Li and Douglas S. Johnson, 17–27 (Hoboken, NJ: Wiley, 2010); Patrick

Dorr and Paul Stupple, "Discovery and Development of Maraviroc and PF-232798: CCR5 Antagonists for the Treatment of HIV-1 Infection," in *Antiviral Drugs*, edited by Wieslaw M. Kazmierski, 117–136 (Hoboken, NJ: Wiley, 2011).

Chapter 3. Hepatitis Viruses

1. N. A. Martin, *J. R. Army Med. Corps* 149 (2003): 121–124.
2. Stephen M. Feinstone, "History of the Discovery of Hepatitis A Virus," *Cold Spring Harbor Perspectives in Medicine* 9, no. 5 (2019): a031740.
3. W. G. E. Cooksley, "What Did We Learn from the Shanghai Hepatitis A Epidemic," *Journal of Viral Hepatitis* 7, suppl. 1 (2000): 1–3.
4. P. J. Provost and M. R. E. Hilleman, "Propagation of Human Hepatitis A Virus In Vitro," *Proceedings of the National Academy of Sciences of the United States of America* 160, no. 2 (1979): 213–221; William H. Bancroft, "Hepatitis A Vaccine," *New England Journal of Medicine* 327, no. 7 (1992): 488–490.
5. Annette Martin and Stanley M. Lemon, "Hepatitis A Virus: From Discovery to Vaccine," *Hepatology* 43, no. 2, suppl. 1 (2006): S164–S172; Bryan Duff and Patrick Duff, "Hepatitis A Vaccine: Ready for Prime Time," *Obstetrics & Gynecology* 91, no. 3 (1998): 468–471; Maria H. Sjorgren, "The Success of Hepatitis A Vaccine," *Gastroenterology* 104, no. 4 (1993): 1214–1216.
6. Ian Gust, "Viral Hepatitis: Some Reflections on a Personal Journey," *Human Vaccines* 5, no. 5 (2009): 357–360.
7. Zhi-Yi Xu and Xuan-Yi Wang, "Live Attenuated Hepatitis A Vaccine Developed in China," *Human Vaccines & Immunotherapeutics* 10, no. 3 (2014): 659–666.
8. Christian Trepo, "A Brief History of Hepatitis Milestones," *Liver International* 34, suppl. 1 (2013): 29–37.
9. Timothy M. Block, Harvey J. Alter, and W. Thomas London, et al., "A Historical Perspective on the Discovery and Elucidation of the Hepatitis B Virus," *Antiviral Research* 131 (2016): 109–123; Baruch M. Blumberg, *Hepatitis B: The Hunt of a Killer Virus* (Princeton, NJ: Princeton University Press, 2003).
10. R. H. Purcell, "The Discovery of the Hepatitis Viruses," *Gastroenterology* 104 (1993): 955–963; Wolfram H. Gerlich, "Medical Virology of Hepatitis B: How It Began and Where We Are Now," *Virology Journal* 10 (2013): 239.
11. Louis Galambos and Jane Eliot Sewell, *Networks of Innovation: Vaccine Development at Merck, Sharp & Dohme, and Mulford, 1895–1995* (Cambridge, UK: Cambridge University Press, 1995); Maurice R. Hilleman, "Vaccines in Historic Evolution and Perspective, a Narrative of Vaccine Discoveries," *Vaccine* 18 (2000): 1438–1447.

12. P. Roy Vagelos and Louis Galambo, *Medicine, Science and Merck* (Cambridge, UK: Cambridge University Press, 2004); P. Roy Vagelos and Louis Galambo, *The Moral Corporation: Merck Experiences* (Cambridge, UK: Cambridge University Press, 2006).

13. Jean Lindenmann, "Interferon and Before," *Journal of Interferon & Cytokine Research* 27, no. 1 (2007): 2–5.

14. Kari Cantell, *The Story of Interferon, The Ups and Downs in the Life of a Scientist* (Singapore: World Scientific, 1998).

15. M. Viganò, G. Grossi, A. Lohlio, et al., *Liver International* 38, suppl. 1 (2018): 79–83.

16. Manoj Kumar and Shiv K. Sarin, "Pharmacology, Clinical Efficacy and Safety of Lamivudine in Hepatitis B Virus Infection," *Expert Review of Gastroenterology & Hepatology* 2, no. 4 (2008): 465–495; Geoffrey Ferir, Suzanne Kaptein, Johan Neyts, et al., "Antiviral Treatment of Chronic Hepatitis B Virus Infections: The Past, The Present and The Future," *Reviews in Medical Virology* 18, no. 1 (2008): 19–34.

17. Emilio Palumbo, "Telbivudine for Chronic Hepatitis B. A Review," *Anti-Infective Agents in Medicinal Chemistry* 7, no. 4 (2008): 245–248.

18. Richard Wilber, Bruce Kreter, Marc Bifano, et al., "Discovery and Development of Entecvir," in *Antiviral Drugs*, edited by Wieslaw M. Kazmierski, 401–416 (Hoboken, NJ: Wiley, 2011); Hong Tang, Jamie Griffin, Steven Innaimo, et al., "The Discovery and Development of a Potent Antiviral Drug, Entecvir, for the Treatment of Chronic Hepatitis B," *Journal of Clinical and Translational Hepatology* 1, no. 1 (2013): 51–58.

19. G. Darby, "The Acyclovir Legacy: Its Contribution to Antiviral Drug Discovery," *Journal of Medical Virology*, suppl. 1 (1993): 134–138; John J. O'Brien and Deborah M. Campoli-Richards, "Acyclovir, An Updated Review of its Antiviral Activity, Pharmacokinetic Properties and Therapeutic Efficacy," *Drugs* 37 (1989): 233–309; Gertrude B. Elion, "The Biochemistry and Mechanism of Action of Acyclovir," *Journal of Microbial Chemotherapy*, suppl. B (1983): 9–17.

20. Erik De Clercq, "Tenofovir: Quo Vadis Anno 2012 (Where Is It Going in the Year 2012)?" *Medicinal Research Reviews* 32, no. 4 (2012): 765–785; Erik De Clercq, "Discovery and Development of Tenofovir Disoproxil Fumarate," in *Antiviral Drugs*, edited by Wieslaw M. Kazmierski, 85–101 (Hoboken, NJ: Wiley, 2011); Erik De Clercq, "An Odyssey in Antiviral Drug Development—50 Years at the Rega Institute: 1964–2014," *Acta Pharmaceutica Sinica B* 5, no. 6 (2015): 520–543.

21. Erik De Clercq, "The Holy Trinity: The Acyclic Nucleoside Phosphonates, Advances in Pharmacology," *Antiviral Agents* 67 (2013): 293–316; Erik De Clercq, "Ten Paths to the Discovery of Antivirally Active Nucleoside and Nucleotide Analogues," *Nucleosides, Nucleotides & Nucleic Acids* 31, no. 4 (2012): 339–352.

22. S. Feng, L. Gao, X. Han, et al., "Discovery of Small Molecule Therapeutics for Treatment of Chronic HBV Infection," *ACS Infectious Diseases* 4 (2018): 257–277.

23. H. J. Alter, "Descartes before the Horse: I Clone, Therefore I Am: The hepatitis C Virus in Current Perspective," *Ann. Intern. Med.* 115, no. 8 (1991): 644–649.

24. M. Houghton, "Discovery of Hepatitis C Virus," *Liver International* 29 (2009): 82–88.

25. T. L. Tellinghuisen, M. J. Evans, T. von Hahn, et al., "Studying Hepatitis C Virus: Making the Best of a Bad Virus," *Journal of Virology* 81, no. 17 (2007): 8853–8867; Stefano Brillanti, Giuseppe Mazzella, and Enrico Roda, "Ribavirin for Chronic Hepatitis C: And the Mystery Goes On," *Digestive and Liver Disease* 43, no. 6 (2011): 425–430.

26. B. D. Lindenbach and C. M. Rice, "Unravelling Hepatitis C Virus Replication from Genome to Function," *Nature* 436, no. 18 (2005): 933–938.

27. Barry Werth, *The Antidote: Inside the World of New Pharma* (New York: Simon & Schuster, 2014).

28. Monste Llinás-Brunet and Peter W. White, "Discovery and Development of BILN 2061 and Follow-up BILN 201335," in *Antiviral Drugs*, edited by Wieslaw M. Kazmierski, 225–238 (Hoboken, NJ: Wiley, 2011).

29. Kevin X. Chen and F. George Njoroge, "The Journey to the Discovery of Boceprevir: An NS3-NS4 HCV Protease Inhibitor for the Treatment of Chronic Hepatitis C," *Progress in Medicinal Chemistry* 49 (2010): 1–36.

30. Barry Werth, *The Antidote: Inside the World of New Pharma* (New York: Simon & Schuster, 2014).

31. Pierre Raboisson, Gregory Fanning, Herman De Kock, et al., "Discovery and Development of the HCV Protease Inhibitor TMC435," in *Antiviral Drugs*, edited by Wieslaw M. Kazmierski, 273–286 (Hoboken, NJ: Wiley, 2011).

32. Nigel J. Liverton, "Evolution of HCV NS3/4a Protease Inhibitors," in *HCV: The Journey from Discovery to a Cure*, edited by M. Sofia, 231–259. Vol. 31 of *Topics in Medicinal Chemistry* (Berlin: Springer, Cham, 2019); James G. Taylor, "Discovery of Voxilaprevir (GS-9857): The Pan-Genotypic Hepatitis C Virus NS3/4A Protease Inhibitor Utilized as a Component of Vosevi," in *HCV: The Journey from Discovery to a Cure*, edited by M. Sofia, 441–457. Vol. 31 of *Topics in Medicinal Chemistry* (Berlin: Springer, Cham, 2019).

33. Nigel J. Liverton, "Evolution of HCV NS3/4a Protease Inhibitors," in *HCV: The Journey from Discovery to a Cure*, edited by M. Sofia, 231–259. Vol. 31 of *Topics in Medicinal Chemistry* (Berlin: Springer, Cham, 2019); James G. Taylor, "Discovery of Voxilaprevir (GS-9857): The Pan-Genotypic Hepatitis C Virus NS3/4A Protease Inhibitor Utilized as a Component

of Vosevi," in *HCV: The Journey from Discovery to a Cure*, edited by M. Sofia, 441–457. Vol. 31 of *Topics in Medicinal Chemistry* (Berlin: Springer, Cham, 2019).

34. Makonen Belema, Nicholas A. Meanwell, John A. Bender, et al., "Discovery and Clinical Validation of HCV Inhibitors Targeting the NS5A Protein," in *Successful Strategies for the Discovery of Antiviral Drugs*, edited by Manoj C. Desai and Nicholas A. Meanwell, 3–28. Vol 32 of *RSC Drug Discovery Series* (Cambridge, UK: RSC Publishing, 2013); Makonen Belema, Steven M. Schnittman, and Nicholas A. Meanwell, "Case History: The Discovery of the First Hepatitis C Virus Replication Complex Inhibitor Daclatasvir (Daklinza)," *Medicinal Chemistry Review 51* (2016): 373–397.

35. John O. Link, "The Discovery of Ledipasvir (GS-5885): The Potent Once-Daily Oral HCV NS5A Inhibitor in the Single-Tablet Regimen Harvoni," in *HCV: The Journey from Discovery to a Cure*, edited by M. Sofia, 57–80. Vol. 32 of *Topics in Medicinal Chemistry* (Berlin: Springer, Cham, 2019); John O. Link, "The Discovery of Velpatasvir (GS-5816): The Potent Pan-Genotypic Once-Daily Oral HCV NS5A Inhibitor in the Single-Tablet Regimens Epclusa and Vosevi," in *HCV: The Journey from Discovery to a Cure*, edited by M. Sofia, 81–110. Vol. 32 of *Topics in Medicinal Chemistry* (Berlin: Springer, Cham, 2019); John O. Link, James G. Taylor, Alejandra Trejo-Martin, et al., "Discovery of Velpatasvir (GS-5816): A Potent Pan-genotypic HCV NS5A Inhibitor in the Single-tablet Regimens Vosevi and Epclusa," *Bioorganic & Medicinal Chemistry Letters 29*, no. 16 (2019): 2415–2427.

36. John O. Link, "The Discovery of Ledipasvir (GS-5885): The Potent Once-Daily Oral HCV NS5A Inhibitor in the Single-Tablet Regimen Harvoni," in *HCV: The Journey from Discovery to a Cure*, edited by M. Sofia, 57–80. Vol. 32 of *Topics in Medicinal Chemistry* (Berlin: Springer, Cham, 2019); John O. Link, "The Discovery of Velpatasvir (GS-5816): The Potent Pan-Genotypic Once-Daily Oral HCV NS5A Inhibitor in the Single-Tablet Regimens Epclusa and Vosevi," in *HCV: The Journey from Discovery to a Cure*, edited by M. Sofia, 81–110. Vol. 32 of *Topics in Medicinal Chemistry* (Berlin: Springer, Cham, 2019); John O. Link, James G. Taylor, Alejandra Trejo-Martin, et al., "Discovery of Velpatasvir (GS-5816): A Potent Pan-genotypic HCV NS5A Inhibitor in the Single-tablet Regimens Vosevi and Epclusa," *Bioorganic & Medicinal Chemistry Letters 29*, no. 16 (2019): 2415–2427.

37. Craig Coburn, "Discovery of Elbasvir," in *HCV: The Journey from Discovery to a Cure*, edited by M. Sofia, 111–131. Vol. 32 of *Topics in Medicinal Chemistry* (Berlin: Springer, Cham, 2019); Michael N. Robertson and Eliav Barr, "Development of ZEPATIER," in *HCV: The Journey from Discovery to a Cure*, edited by M. Sofia, 369–407. Vol. 32 of *Topics in Medicinal Chemistry* (Berlin: Springer, Cham, 2019).

38. Rolf Wagner, David A. DeGoey, John T. Randolph, et al., "HCV NS5A as an Antiviral Therapeutic Target: From Validation to the Discovery and Development of Ombitasvir and Pibrentasvir as Components of IFN-Sparing HCV Curative Treatments," in *Successful Strategies for the Discovery of Antiviral Drugs*, edited by Manoj C. Desai and Nicholas A. Meanwell, 133–156. Vol 32 of *RSC Drug Discovery Series* (Cambridge, UK: RSC Publishing, 2013).

39. Aesop Cho, "Evolution of HCV NS5B Nucleoside and Nucleotide Inhibitors," in *HCV: The Journey from Discovery to a Cure*, edited by M. Sofia, 117–139. Vol. 31 of *Topics in Medicinal Chemistry* (Berlin: Springer, Cham, 2019).

40. Michael J. Sofia and Phillip A. Furman, "The Discovery of Sofosbuvir: A Liver- Targeted Nucleotide Prodrug for the Treatment and Cure of HCV," in *HCV: The Journey from Discovery to a Cure*, edited by M. Sofia, 141–169. Vol. 31 of *Topics in Medicinal Chemistry* (Berlin: Springer, Cham, 2019); Ivan Gentile, Alberto Enrico Maraolo, Antonio Riccardo Buonomo, et al., "The Discovery of Sofosbuvir: A Revolution for Therapy of Chronic Hepatitis C," *Expert Opinion on Drug Discovery* 10, no. 12 (2015): 1363–1377.

41. Barry Werth, *The Antidote: Inside the World of New Pharma* (New York: Simon & Schuster, 2014).

42. Peter Palese, "Profile of Charles M. Rice, Ralf F. W. Bartenschlager, and Michael J. Sofia, 2016 Lasker-DeBakey Clinical Medical Research Awardees," *Proceedings of the National Academy of Sciences of the United States of America* 113, no. 49 (2016): 13934–13937.

43. William J. Watkins, "Evolution of HCV NS5B Non-nucleoside Inhibitors," in *HCV: The Journey from Discovery to a Cure*, edited by M. Sofia, 171–191. Vol. 31 of *Topics in Medicinal Chemistry* (Berlin: Springer, Cham, 2019).

44. Robert G. Gentles, "Discovery of Beclabuvir: A Potent Allosteric Inhibitor of the Hepatitis C Virus Polymerase," in *HCV: The Journey from Discovery to a Cure*, edited by M. Sofia, 193–228. Vol. 31 of *Topics in Medicinal Chemistry* (Berlin: Springer, Cham, 2019).

Chapter 4. Influenza: A Perennial Killer

1. Darwyn Kobasa and Yoshihiro Kawaoka, "Emerging Influenza Viruses: Past and Present," *Current Molecular Medicine* 5 (2005): 791–803.

2. K. Kuszewski and L. Brydak, "The Epidemiology and History of Influenza," *Biomedicine & Pharmacotherapy* 54, no. 4 (2000): 188–195.

3. Claude Hannoun, "The Evolving History of Influenza Viruses and Influenza Vaccines," *Expert Rev. Vaccines* 12, no. 9 (2013): 1085–1094.

4. J. S. Oxford, "Influenza A Pandemics of the 20th Century with Special Reference to 1918: Virology, Pathology and Epidemiology," *Reviews in Medical Virology* 10, no. 2 (2000): 119–1133.

5. Alfred W. Crosby, *America's Forgotten Pandemic, The Influenza of 1918* (Cambridge, UK: Cambridge University Press, 1989).

6. John M. Barry, *The Great Influenza: The Story of the Deadliest Pandemic in History* (New York: Penguin Books, 2004).

7. Gina Kolata, *Flu: The Story of The Great Influenza Pandemic of 1918 and the Search for the Virus that Caused It* (New York: Touchstone, 2001).

8. Gina Kolata, *Flu: The Story of The Great Influenza Pandemic of 1918 and the Search for the Virus that Caused It* (New York: Touchstone, 2001).

9. Jeffrey K. Taubenberger and John C. Kash, "Insights on Influenza Pathogenesis from the Grave," *Virus Research* 162 (2011): 2–7.

10. Zachary Cope, *Almroth Wright, Founder of Modern Vaccine-therapy.* (London, UK: Thomas Nelson Ltd, 1966).

11. P. E. Castle and M. Maza, "Prophylactic HPV Vaccination: Past, Present, and Future," *Epidemiology and Infection* 144, no. 3 (2016): 449–468.

12. Stanley Plotkin, "History of Vaccination," *Proceedings of the National Academy of Sciences of the United States of America* 111, no. 34 (2014): 12283–12287; Stanley A. Plotkin and Susan L. Plotkin, "The Development of Vaccines: How the Past Led to the Future," *Nature Reviews Microbiology* 9, no. 12 (2011): 889–893.

13. John Treanor, "History of Live, Attenuated Influenza Vaccine," *Journal of the Pediatric Infectious Diseases Society* 9, suppl. 1 (2020): S3–S9.

14. S. M. Wintermeyer and M. C. Nahata, "Rimantadine: A Clinical Perspective," *Ann. Pharmacother.* 29 (1995): 299–310; P. H. Jalily, M. C. Duncan, D. Fedida, et al., "Put a Cork in It: Plugging the M2 Viral Ion Channel to Sink Influenza," *Antiviral Res.* 178 (2020): 104780.

15. G. Hubsher, M. Haider, and M. S. Okun, "Amantadine: The Journey from Fighting Flu to Treating Parkinson Disease," *Neurol.* 78 (2012): 1096–1099.

16. William Graeme Laver, "From the Great Barrier Reef to a 'Cure' For the Flu: Tall Tales, But True," *Perspectives in Biology and Medicine* 47, no. 4 (2004): 590–596.

17. J. N. Varghese, J. L. McKimm-Breschkin, J. B. Caldwell, et al., "The Structure of the Complex between Influenza Virus Neuraminidase and Sialic Acid, the Viral Receptor," *Proteins* 14 (1992): 327–332.

18. M. von Itzstein, "The War Against Influenza: Discovery and Development of Sialidase Inhibitors," *Nat. Rev. Drug Discov.* 6 (2007): 967–974; R. Thomson and M. von Itzstein, "Discovery and Development of Influenza Virus Sialidase Inhibitor Relenza," in *Antiviral Drugs*, edited by Wieslaw M. Kazmierski, 385–400 (Hoboken, NJ: Wiley, 2011).

19. M. von Itzstein, "The War Against Influenza: Discovery and Development of Sialidase Inhibitors," *Nat. Rev. Drug Discov.* 6 (2007): 967–974; R.

Thomson and M. von Itzstein, "Discovery and Development of Influenza Virus Sialidase Inhibitor Relenza," in *Antiviral Drugs*, edited by Wieslaw M. Kazmierski, 385–400 (Hoboken, NJ: Wiley, 2011).

20. W. Lew, X. Chen, and C. K. Kim, "Discovery and Development of GS 4104 (Oseltamivir): An Orally Active Influenza Neuraminidase Inhibitor," *Curr. Med. Chem.* 7 (2000): 663–672.

21. N. A. Meanwell, S. V. D'Andea, C. W. Cianci, et al., "Antiviral Drug Discovery," in *Drug Discovery: Practices, Processes, and Perspectives*, edited by J. J. Li and E. J. Corey, 439–515 (Hoboken, NJ: Wiley, 2013).

22. N. A. Meanwell, S. V. D'Andea, C. W. Cianci, et al., "Antiviral Drug Discovery," in *Drug Discovery: Practices, Processes, and Perspectives*, edited by J. J. Li and E. J. Corey, 439–515 (Hoboken, NJ: Wiley, 2013).

23. C. Sweet, K. J. Jakeman, K. Bush, et al., "Oral Administration of Cyclopentane Neuraminidase Inhibitors Protects Ferrets Against Influenza Virus Infection," *Antimicrob. Agents Chemother.* 46 (2002): 996–1004.

24. M. M. McLaughlin, E. W. Skoglund, and M. G. Ison, "Peramivir: An Intravenous Neuraminidase Inhibitor," *Exp. Opin. Pharmacother.* 16 (2015): 1889–1900.

25. S. Kubo, T. Tomozawa, M. Kakuta, et al., "Laninamivir Prodrug CS-8958, A Long-acting Neuraminidase Inhibitor, Shows Superior Anti-influenza Virus Activity After a Single Administration," *Antimicrob. Agents Chemother.* 54 (2010): 1256–1264.

26. M. Shirley, "Baloxavir Marboxil: A Review in Acute Uncomplicated Influenza," *Drugs* 80 (2020): 1109–1118.

Chapter 5. Coronaviruses

1. A. Zumla, J. F. W. Chan, E. I. Azgar, et al., "Coronavirus—Drug Discovery and Therapeutic Options," *National Review of Drug Discovery* 15, no. 5 (2016): 327–347; D. S. Hui, G. A. Rossi, and S. L. Johnston, eds., *SARS, MERS and other Viral Lung Infections* (Sheffield, UK: European Respiratory Society, 2016).

2. V. G. da Costa, M. L. Moreli, and M. V. Saivish, "The Emergence of SARS, MERS and Novel SARS-2 Coronaviruses in the 21st Century," *Arch. Virol.* 165, no. 7 (2020): 1517–1526; I. R. Parrey and P. R. De los Rios-Escalante, "Novel Coronavirus (COVID-19) History, Genome Structure and Life Cycle—A Review," *Chem. Biol. Interface* 11, no. 1 (2021): 1–5.

3. N. Zhu, D. Zhang, W. Wang, et al., "A Novel Coronavirus from Patients with Pneumonia in China," *New. Engl. J. Med.* 382 (2020): 727–733.

4. N. B. Mercado, R. Zahn, F. Wegmann, et al., "Single-shot Ad26 Vaccine Protects Against SARS-CoV-2 in Rhesus Macaques," *Nature* 586, no. 7830 (2020): 583–588.

5. N. Tatsis and C. J. H. Ertl, "Adenoviruses as Vaccine Vectors," *Molecular Therapy* 10, no. 4 (2004): 616–629.

6. J. Custers, et al., "Vaccines Based on Replication Incompetent Ad26 Viral Vectors: Standardized Template with Key Considerations for a Risk/Benefit Assessment," *Vaccines* 39 (2021): 3081–3101.

7. M. J. Sofia, D. Bao, W. Chang, et al., "Discovery of a β-D-2'-Deoxy-2'-α-fluoro-2'-β-C-methyluridine Nucleotide Prodrug (PSI-7977) for the Treatment of Hepatitis C Virus," *J. Med. Chem.* 53 (2010): 7202–7218.

8. A. Cho, O. L. Saunders, T. Butler, et al., "Synthesis and Antiviral Activity of a Series of 1'-substituted 4-aza-7,9-dideazaadenosine C-nucleosides," *Bioorg. Med. Chem. Lett.* 22 (2012): 2705–2707.

9. A. Cho, L. Zhang, J. Xu, et al., "Discovery of the First *C*-Nucleoside HCV Polymerase Inhibitor (GS-6620) with Demonstrated Antiviral Response in HCV Infected Patients," *J. Med. Chem.* 57 (2014): 1812–1825.

10. D. Siegel, H. C. Hui, E. Doerffler, et al., "Discovery and Synthesis of a Phosphoramidate Prodrug of a Pyrrolo[2,1-*f*][triazin-4-amino] Adenine *C*-Nucleoside (GS-5734) for the Treatment of Ebola and Emerging Viruses," *J. Med. Chem.* 60 (2017): 1648–1661.

11. T. K. Warren, V. Soloveva, J. Wells, et al., "Therapeutic Efficacy of the Small Molecule GS-5734 against Ebola Virus in Rhesus Monkeys," *Nature* 531 (2016): 381–385.

12. E. Murakami, T. Tolstykh, H. Bao, et al., "Mechanism of Activation of PSI-7851 and Its Diastereoisomer PSI-7977," *J. Biol. Chem.* 285 (2010): 34337–34347; H. Ma, W. -R. Jiang, N. Robledo, et al., "Characterization of the Metabolic Activation of Hepatitis C Virus Nucleoside Inhibitor β-d-2'-Deoxy-2'-fluoro-2'-C-methylcytidine (PSI-6130) and Identification of a Novel Active 5'-Triphosphate Species," *J. Biol. Chem.* 282 (2007): 29812–29820.

13. T. K. Warren, V. Soloveva, J. Wells, et al., "Therapeutic Efficacy of the Small Molecule GS-5734 against Ebola Virus in Rhesus Monkeys," *Nature* 531 (2016): 381–385.

14. E. Murakami, T. Tolstykh, H. Bao, et al., "Mechanism of Activation of PSI-7851 and Its Diastereoisomer PSI-7977," *J. Biol. Chem.* 285 (2010): 34337–34347; H. Ma, W. -R. Jiang, N. Robledo, et al., "Characterization of the Metabolic Activation of Hepatitis C Virus Nucleoside Inhibitor β-d-2'-Deoxy-2'-fluoro-2'-C-methylcytidine (PSI-6130) and Identification of a Novel Active 5'-Triphosphate Species," *J. Biol. Chem.* 282 (2007): 29812–29820.

15. R. T. Eastman, J. S. Roth, K. R. Brimacombe, et al., "Remdesivir: A Review of Its Discovery and Development Leading to Emergency Use Authorization for Treatment of COVID-19," *ACS Cent. Sci.* 6, no. 5 (2020): 672–683.

16. G. E. Arth, J. Fried, D. B. R. Johnston, et al., "16-Methylated Steroids. II. 16α-methyl Analogs of Cortisone, A New Group of Anti-Inflammatory Steroids. 9α-halo Derivatives," *J. Am. Chem. Soc.* 80 (1958): 3161–3163.

17. R. Hirschmann, "The Cortisone Era: Aspects of Its Impact. Some Contributions of the Merck Laboratories," *Steroids* 57 (1992): 579–592.

18. J. Fried and E. F. Sabo, "9α-fluoro Derivatives of Cortisone and Hydrocortisone," *J. Am. Chem. Soc.* 76 (1954): 1455–1456.

19. H. Herzog and E. P. Oliveto, "A History of Significant Steroid Discoveries and Developments Originating at the Schering Corporation (USA) since 1948," *Steroids* 57 (1992): 617–623.

Chapter 6. Closing Remarks

1. David M. Morens and Jeffery K. Taubenberger, "Influenza Cataclysm, 1918," *The New England Journal of Medicine* 379, no. 24 (2018): 2285–2287.

Index

For the benefit of digital users, indexed terms that span two pages (e.g., 52–53) may, on occasion, appear on only one of those pages.

3TC, 62, 63, 116, 125, 204
3TC, structure, 220, 224
6′-deoxychalcone, 5, 78, 87, 134, 139

abacavir, 62, 65, 67, 204
abacavir, structure, 220
ACE2, 190, 191, 196
acquired immunodeficiency syndrome, 25
acquired T-cell defect, 38
actinomycin D, 34
active pharmaceutical ingredient, 73
acyclic nucleotide, 65, 67, 120–122
acyclic nucleotide phosphonates, 120
acyclovir, 120, 121, 203
Ad26, 196, 199, 200, 201
Ad5, 199, 200
adamantane, 172
adaptive immune response, 38, 194, 198
Addison's disease, 212
adefovir dipivoxil, 123, 124
adefovir dipivoxil structure, 224
adenine 57, 57
adenovirus serotype 26, 199, 200
adenoviral vectors, 199
adenovirus, 40, 197–201
adult T-cell leukemia, 35, 36
Agenerase, 135
agglutination, 102, 156

agglutinogen, 102
agonist, 215
AIDS vaccines, 84
alafenamide, 83, 117, 120, 124, 125, 223, 225
alafenamide structure, 225
alanine aminotransferase, 117, 123, 125
Alexander cells, 109
ALT, 117, 123, 125
aluminum hydroxide adjuvant, 97, 98, 110
amantadine, 172, 173, 174, 175, 231
amantadine structure, 231
amide hydrolysis, 75
amniotic fluid, 168
amprenavir, 77, 135
amprenavir structure, 222
amyl nitrate, 39
angiotensin converting enzyme-2, 190
animal reservoirs, 188
anthocyanins, 5
anthrax bacillus, 21
anticoagulant, 102
antigen particles, 107
antigen shift, 157, 165, 171
antigen–antibody reaction, 101, 102, 152, 156
antigen-encoded mRNA vaccines, 195
antigenic change, 170

antigenically different viruses, 172
antigenicity, 107
antigen-presenting cells, 198
anti-inflammatory steroid, 211
anti-interferon serum, 44
anti-muscarinic, 66
antitoxin, 166
API, 73
aprepitant, 205
aptivus, 77, 79
arlansa, 134
asparagine, 68
aspartic proteases, 70
asunaprevir, 138
asunaprevir structure, 227
atazanavir structure, 222
attenuated influenza A, 171
attenuated rabies virus, 166
attenuated vaccines, 167
Australian antigen, 30, 101, 103, 104, 107
avian flu, 155, 173, 217
avian leukemia virus, 25
Avigan, 184
azidothymidine, 40
AZT, 40, 56, 58, 60–63, 76, 89, 203
Aztec empire, 5–8
AZT-triphosphate, 60

backbone binding, 79
bakers' yeast, 109, 110, 167
balapiravir, 144
baloxavir, 185
baloxavir acid, 185
baloxavir acid structure, 232
baloxavir marboxil, 184–186
baloxavir marboxil structure, 232
Baltimore, David, 34, 35, 41
Banting, Frederick, 52, 53, 128
Baraclude, 118, 120
Barré-Sinoussi, Françoise, 40–52
B-cells, 38, 95, 197
BCH-18963, 64

BCS, 86
beclabuvir, 150
beclabuvir structure, 230
Berkefeld filter, 26
beta-lipoproteins, 100
betamethasone, 214
bictegravir, 83
bictegravir structure, 223
Biktarvy, 83
biliary clearance, 72
binding pocket, 68, 69, 85, 131, 137, 182
bioavailability booster, 73
Biopharmaceutics Classification System, 86
blood transfusion, 102
Blumberg, Baruch, 30, 93, 99–111, 127
boceprevir, 133, 134, 136
boceprevir structure, 226
booster shot, 199
brivudine, 58
bromovinyldeoxyuridine, 58
bromovinyldeoxyuridine structure, 219
budding, 175
Burkitt's lymphoma, 27, 28
bursa of Fabricius, 38
butterfly-like shape, 69

Cabenuva, 90
cabotegravir, 90, 223
cabotegravir structure, 223
cancer etiology, 31
cancer-causing viruses, 25–32
cap-dependent endonuclease inhibitor, 184–186
capsid, 25, 65, 167
carbocyclic deoxyguanosine analog, 116
carbocyclic nucleosides, 65
carboxylesterase 1, 210
cardiotoxicities, 87, 132

casein, 114
CatA1, 210
catalytic triad, 138, 144
cathepsin A 1, 210
cathepsins, 70
Catherine the Great, 14
CCR5 antagonist, 83, 87, 88
CCR5 inhibitors, 86–88
CD4 cell, 84
CD4 counts, 61
CD4 receptor, 84
CD4, 46, 61, 70, 84, 86
CD4-positive T-cells, 46
CDC, 39, 46, 128, 169, 174, 207
cDNA, 32
cell-based antiviral screens, 206
cellular kinases, 119
cellular protease homologue, 81
Center for Disease Control and
    Prevention, 39, 46, 128, 169,
    174, 207
central dogma, 33, 38, 194
central nervous system, 173
Cervarix, 31
cervical cancer, 30
CES1, 210
chain terminator, 61, 66, 121, 123
Chamberland bougies, 26
Chamberland filter, 22, 23
Chamberland filter-candle, 22
charge–charge interactions, 177
chelating ligand, 83, 185
chemokine receptors, 86
chemotype, 132, 136, 140, 149
chick embryo cells, 114, 164, 168
chicken embryo cultures, 161
chimpanzee, 37, 129, 199, 200
chimpanzee adenovirus, 200
chiral resolution, 63
cholera bacillus, 21
chorioallantoic membrane, 113
chronic obstructive pulmonary
    disease, 215

ciluprevir, 132, 136
ciluprevir structure, 226
cirrhosis, 99, 126
clevudine, 145
clinically silent, 125
cloning, 109, 110, 115, 128, 152
Clusius, Carolus, 2
cluster of differentiation, 46
CMV, 119
CNS, 173
C-nucleoside, 206, 209
cobicistat, 73, 75, 80
cocktail drugs, 40, 72
combination drug therapy, 63
Comirnaty, 192
complementary DNA, 32
complementary mechanisms, 138
composition of matter patent, 59
computer-aided drug design, 76, 177
conquistadores, 6
conserved catalytic site, 176
COPD, 215
coronaviral genome, 188
Coronaviridae family, 188, 207
coronavirus disease-2019, 187
Cortés, Hernán, 5
corticosteroids, 212
cortisol, 212
cortisone, 211, 212
cortisone era, 212
covalent closed circular DNA, 106
cowpox, 7
CRISPR-Cas9, 87
Crixivan, 75
Crohn's disease, 211
cross-resistance, 185
crystalline form, 73
CTCL, 35, 42
cutaneous T-cell lymphoma, 35, 42
CXCR4, 86
cyclohexene, 180
cyclopentane ring, 182
cyclopropyl acylsulfonamide, 137

cyclosporine, 39
CYP2D6 inhibition, 88
CYP3A4, 72, 74, 75
CYP450 enzymes, 149, 214
cyromegalovirus, 119
cytochrome P450 3A4, 72
cytokine, 113, 115
cytopathic effect, 96, 97
cytoplasm, 196
cytosine 57, 57
cytotoxicity, 58, 63, 66, 203, 207
Cytovene, 121, 203

d4T, 58, 59, 62, 204, 220
d4T structure, 220
daclatasvir, 139, 140, 143, 150, 228
daclatasvir structure, 228, 231
Daklinza, 140, 143, 228
DANA, 177, 178, 180, 231
DANA structure, 231
Dane particles, 104
danoprevir, 138
danoprevir structure, 227
darunavir, 79
darunavir structure, 222
dasabuvir structure, 230
dasabuvir, 150
daughter viruses, 156, 163, 174
ddC, 62, 63
DDI, 75
De Clercq, Erik, 58, 68, 121–123
Decadron, 211–216
delavirdine, 67
delavirdine structure, 221
Denavi, 121, 203
dengue, 207
deoxynucleotide triphosphates, 34
deoxyribonucleic acid, 4, 38
dexamethasone, 211–216
dexamethasone structure, 233
didanosine structure, 220
diffraction-grade crystals, 135
diketoacid derivative, 82, 185

dipeptidyl-peptidase 4, 190
diphtheria vaccine, 166
DNA chain terminators, 61, 117, 118, 121, 123
DNA plasmid vaccines, 197
DNA polymerase, 35, 58, 60, 106, 107, 118, 203
DNA provirus hypothesis, 33
DNA retrovirus, 105
DNA tumor virus, 29
D-nucleosides, 118
dolutegravir, 83, 185
dolutegravir structure, 223
doravirine, 68, 69
doravirine structure, 221
double-stranded RNA, 43, 113, 198, 201
double-stranded viral DNA, 81
Down syndrome, 103, 104
DPP-4, 190
drug–drug interaction, 75
drug-resistant strains, 62
dsDNA, 198
Dumas, Alexandre, 3

E. coli, 109
Ebola virus, 200, 207, 209
EBOV, 200, 207, 209
Edurant, 68, 69
efavirenz, 67
efavirenz structure, 221
EGFR, 41
egg white, 168
elbasvir, 140, 142
elbasvir structure, 229
electron micrograph, 1, 24, 42, 45, 94, 103
electrophilic trap, 133, 134
ELISA, 128, 192
ELISA, 45, 128, 192
Elpida, 69
elsufavirine, 69
elvitegravir, 74, 82

elvitegravir structure, 223
EMA, 200
embryonated eggs, 169
emend, 205
emergency use authorization,
    183, 192
Emperor Shunzhi, 12
emtricitabine, 62, 64, 65, 83, 145, 220
emtricitabine structure, 220
enantiomers, 63, 64, 118
endemic, hepatitis B, 99
endogenous ligand, 137
endo-nucleolytic "cap-snatching"
    reaction, 184
enfuvirtide, 83, 85
entecavir, 118, 120
entecavir structure, 224
enzyme-linked immunosorbent
    assay, 45, 128, 192
epidermal growth factor
    receptor, 41
epitope-mapping, 85
epitopes, 128
Epivir, 62, 63, 76, 116, 117, 125, 204
Epstein–Barr virus, 27, 28, 47
equine anemia virus, 45
erlotinib, 41
esterases, 123
ethyldeoxyuridine structure, 219
etiologic agent, 161
etravirine, 68, 69
etravirine structure, 221
EUA, 183, 192, 193, 196
eukaryotic cells, 194
European Medicines Agency, 200
excipient-to-drug, 77

factor VIII, 98
faldaprevir, 132, 136, 184
faldaprevir structure, 226
fecal–oral route transmission, 93
feline infectious peritonitis, 188
ferrets, 155, 162

fertilized chick embryo
    technique, 164
fetal rhesus kidney cell line, 97
Fibiger, Johannes, 54, 55
fibroblast interferon, 115
filibuvir, 149
Filoviridae family, 207
filterable agent, 18, 22, 25, 26, 161
fingers subdomains,
    polymerases, 144
Finlay, Cárlos, 18
FIP, 61
first-generation HIV protease
    inhibitors, 71–77
first-to-patient, 61
flavivrus family, 128
flavonoids, 5
fludrocortisone, 205, 212
Flumadine, 174
fluorination, 64
foot-and-mouth disease virus 23, 43
formalin-inactivated virus, 97, 98
fosamprenavir, 77
fosamprenavir structure, 222
fostemsavir, 86
French and Indian war, 10
Friend virus, 60
FTC, 62, 64, 145
functional proteins, 131
Fuzeon, 83, 85

Gallo, Robert, 35, 40–52, 55, 59,
    68, 115
ganciclovir, 121, 123, 203
Ganovo, 138
Gardasil, 31, 167
gastricsin, 70
Gastrozepin, 66
gay-related immunodeficiency 39
gefitinib, 41
gene reassortment, 157
gene splicing technique, 115, 116
genetic cargo, 198

genetic drift, 189
genetic payload, 197
genetic splicing, 109
Genghis Khan, 7
genome sequence of
    SARS-CoV-2, 193
germ theory, 19–22
Ghosh, Arun, 79
glecaprevir structure, 228
glecaprevir, 138
glucocorticoids, 212
glycoprotein, 33, 65, 70, 74, 84, 131,
    156, 170, 171, 175, 191, 193, 194, 196
glycoprotein envelope, 33
gold standard of clinical trials, 209
gp, 65, 70, 74, 83, 84, 85, 86
gp120, 65, 70, 84, 85, 86
gp160, 85
gp41, 70, 83, 84, 85
GPCR, 86
G-protein coupled receptors, 86
grazoprevir, 138, 142
grazoprevir structure, 227
GRID, 39
guanidine group, 57, 57, 178, 182
guanosine, 120
Guillain–Barre disease, 169

H1N1 virus, 165, 171, 181, 182
H274Y mutation, 183
H5N1 influenza A virus, 181, 217
H5N1 swine flu, 169
HA, 156
HAART, 40
Harvoni, 141
HAV antigens, 94
HAV, 93–98
Havrix, 95
HBV, 98–125
HBV genome, 106
Helicobacter pylori, 179
hemagglutinin, 44, 156, 157, 164,
    169–172, 174, 175, 191

hemoglobin, 102, 156
hemophiliacs, 39, 40, 42, 98, 101
Hemophilius influenza, 159
Hepadnaviridae family, 106
hepadnavirus, 30
hepatitis A vaccines, 95–98
hepatitis A virus, 93–98
hepatitis B drugs, 112–125
hepatitis B specific antiviral drug, 117
hepatitis B surface antigen, 101
hepatitis B vaccine, 43, 106–111
hepatitis B virus, 30, 98–125, 217
hepatitis C virus discovery, 53
hepatitis C virus NS3/4A serine
    protease inhibitors, 130–139
hepatocellular carcinoma, 30, 99,
    110, 126
hepatocytes, 106, 118, 146
hepatoxicity, 138
Hepsera, 124
Heptavax-B, 108, 109, 110
herd immunity, 195, 201
hERG, 87
herpes simplex keratitis, 57
herpes simplex virus, 57, 58, 119,
    121, 203
herpesvirus-encoded thymidine
    kinase, 121
hexapeptide, 135
high throughput screen, 78, 87,
    134, 139
highly active antiretroviral
    therapy, 40
Hilleman, Maurice, 97, 108, 113, 114
Hint 1, 210
HIV antibodies, 48
HIV blood test, 48
HIV entry inhibitors, 83–88
HIV fusion inhibitors, 84–88
HIV integrase inhibitors, 80–83
HIV protease inhibitors, 69–80
HIV structure, 65
HLTV-IIIB, 47, 48

Holý Trinity, 120–125
Holý, Antonín, 121–123
horseshoe-like conformation, 69
host polymerase, 118
host-cell nuclei, 197
HPV vaccine, 31
HPV, 30, 51, 167, 119, 203
HSV, 57, 58, 119, 121, 203
HTLV, 35, 45, 46, 49
HTS hit, 142
HTS, 78, 87, 88, 134, 139, 142
human ether-a-go-go-related gene, 87
human immunodeficiency virus, 35
human lymphocytic cells, 29
human papilloma viruses, 30, 51, 167
human T-cell leukemia virus, 35
hydrocortisone, 212
hydrophobic contacts, 177
hydrophobic pocket, 180, 182

idoxuridine, 56–58, 202, 219
idoxuridine structure, 219
IdU, 56–58, 202, 219
Ig, 95
IgG immunoglobins, 192
IgM immunoglobins, 192
immune electron microscope,
    94, 152
immune response, 38, 101, 192, 194,
    197, 198, 199
immunity, 7, 8, 11, 12, 17, 160, 168,
    171, 192, 195, 199, 200, 202,
    211, 218
immunocompetent, 178
immunocompromised persons, 194
immunodiffusion assay, 105
immunogenicity, 97, 109, 111,
    167, 198
immunoglobin, 95, 96, 192
inactivated vaccines, 167
inactivated virus vaccines, 197
inavir, 183–184
Inca Empire, 9–10

Incivek, 134
IND, 60, 61
indinavir, 75
indinavir structure, 221
induced fit, 177
infectious hepatitis, 92
infective hepatitis, 98
influenza epidemics, 155
influenza virus structure, 172
innate immune responses, 198
INSTI, 74, 80–83, 90
insulin discovery, 52–53
integrase strand transfer inhibitor,
    74, 80–83, 90
Intelence, 68, 69
interferon, 43, 44, 112, 113–116, 129,
    130, 133, 138, 148, 149
interleukin-2, 41, 43, 44, 115
interleukin, 41, 43, 44, 113, 115
intravenous, 109, 126, 182
intrinsic hypothesis, 25
Intron, 115
Investigational New Drug
    Application, 60, 61
Invirase, 71
Iopinavir structure, 222
Iressa, 41
Isentress, 82
islatravir, 64
isoleucine, 68, 118
Ivanovsky, Dmitry, 23
ivermectin, 110

Janssen, Paul, 68
Jcovden, 192

Kangxi, 12
Kaposi's sarcoma, 39
ketoamides, 134, 135
killed-virus vaccine, 167
kinase bypass, 122, 145, 146, 147
kinases, 42, 58, 61, 118, 119, 121, 122,
    145, 146, 147, 203, 206, 210

Koch's four postulates, 21, 22, 104
Koch's postulates, 21, 22, 104
KRAS-G12C, 55

lamivudine, 62–65, 76, 116–120, 123,
    125, 204
lamivudine structure, 220, 224
Landsteiner, Karl, 101, 102
laninamivir, 183–184
laninamivir octanoate structure, 232
Laninamivir octanoate, 183, 232
laninamivir structure, 232
LAV, 46
ledipasvir, 140, 141
ledipasvir structure, 228
lentiretrovirus, 45
lentivirus 45
leucine, 67
leukemia, 25, 26, 35, 36, 41, 42, 43,
    60, 103, 116
leukocyte, 114, 115
leukocyte interferon, 114, 115
Lexiva, 77
lipid envelope, 129, 163
lipid nanoparticle, 195
Lipinski's rule-of-five, 151, 152
liposomal nanoparticle, 194
live influenza vaccine, 168
liver toxicities, 67
liver transplantation, 126
LNP, 195
L-nucleosides, 116, 118
lobucavir, 119
Loeffler, Friedrich, 23, 24
lomibuvir, 149
long-acting HIV/AIDS drugs, 90
Louisiana Purchase 15–18
lymph gland, 43
lymph node cells, 44
lymphadenopathy, 44
lymphadenopathy-associated
    virus, 46
lymphocytes, 28, 35, 42, 44, 46,
    60, 115

lymphoma, 27, 28, 29, 35, 42
lysine, 68

M2 inhibitors, 172–175
M2 ion channel protein, 164,
    174, 175
magnesium divalent ion, 82, 83, 184
major capsid antigen L1 protein, 167
maraviroc, 83, 88
maraviroc structure, 223
marmoset monkeys, 97
matrix M1 protein, 164, 174
Mavyret, 143
mechanism of action, 58, 61, 85, 142,
    174, 203, 209
mechanism-based cardiotoxicity, 132
Mectizan, 110
Medawar, Peter B.1, 33
memory T-cells, 89, 166, 201
mericitabine, 146
MERS, 189–190, 207, 208
messenger ribonucleic acid, 33, 194
messenger RNA, 115, 198
metal-chelating functionalities, 185
method of use patent, 59
miasmatic theory, 19
Middle East respiratory syndrome,
    189–190, 207, 208
mineralocorticoids, 212
mitochondrial DNA, 66
mitochondrial toxicity, 58, 203
mixed vessel, 155
MoA, 203, 209, 210, 215
molecular biology, 35, 41, 75, 127,
    128, 162
molecular peanut butter, 72
Montagnier, Luc, 40–52, 68, 127
mRNA, 33, 34, 185, 192–197, 198,
    201, 202
mRNA vaccines, 192–197, 201, 202

NA, 156, 172
nanotechnology, 195
narlaprevir, 134

narlaprevir structure, 226
nasopharyngeal inoculation, 161
NDA, 88
NDPK, 210
nelfinavir, 76
nelfinavir structure, 222
nesbuvir, 149
neuraminic acid, 156, 177
neuraminidase, 156, 157, 163, 164,
   170, 171, 174–185, 232
neuraminidase inhibitors, 175–184,
   231, 232
neurokinin 1 inhibitor, 170, 205
neutralizing antibodies, 196, 201
nevirapine, 66, 67, 221
nevirapine structure, 221
New Drug Application 60, 88
NK1 inhibitor, 170, 205
NNRTI, 66, 204
N-nucleosides, 209
non-A, non-B viral hepatitis, 127
non-biodegradable, 121
non-cleavable transition-state
   mimetic, 77
non-functional polypeptides, 131
non-human adenovirus, 200
non-nucleoside HCV NS5B
   inhibitors, 148–150
nonnucleoside inhibitors, 144
non-nucleoside reverse transcriptase
   inhibitors, 65–69, 90
non-steroidal anti-inflammatory
   drugs, 215
non-structure proteins, 130, 203
Norvir, 72, 73
NRTI, 61–65, 204
NS proteins, 130, 203
NS3 serine protease, 130, 131, 133
NS5A protein inhibitor, 139–143
NS5B polymerase inhibitors,
   143–150
NS5B polymerase, 139, 143–150
NSAIDS, 215
nucleocapsid, 188, 191, 192

nucleocapsid antigens, 192
nucleoprotein, 164, 172, 174
nucleoside/tide inhibitors, 144–148
nucleoside antiviral drugs for
   AIDS, 57, 57
nucleoside diphosphate kinase, 210
nucleoside reverse transcriptase
   inhibitor, 61–65, 204
nucleoside triphosphate, 203
nucleotide, 32, 34, 58, 60, 65, 81, 89,
   117, 118, 120–123, 125, 144–146,
   148, 165, 194, 203, 205–207, 210
nucleotide diphosphate, 58, 203
nucleotide monophosphate, 58, 203
nucleotide triphosphate, 58, 203

olfactory function, 158
Olysio, 137
ombitasvir, 140, 142, 150, 229
ombitasvir structure, 229
oncogene, 27, 31, 55
oncogenic genome, 167
original antigenic sin, 165
orthomyxovirus family, 163
orthopoxvirus, 7
oseltamivir, 179–181, 232
oseltamivir structure, 232
oseltamivir-resistant influenza
   virus, 183
osimertinib, 41–42
Ouchterlony double
   imminodiffusion test, 100
oxathiolane core structure, 63, 117
oxetanocin A, 119
oxonium ion, 177, 180

PA, 184, 185
palm subdomains, polymerases, 144
palm-II inhibitors, 148
parenteral formulation, 182
paritaprevir, 142, 143, 150, 227
paritaprevir structure, 227
Parkinson's disease, 173, 174
parvovirus contaminant, 97

passive double immunodiffusion, 100
Pasteur, Louis, 20, 21, 22, 37, 166
patent busting, 140, 150
pathogenic, 165, 181, 189, 208, 217
PB1, 184
PB2, 184
PCR, 164, 192
pegylated interferon, 116, 129, 133, 138, 149
pegylated interferon-α, 129
pegylated interferon-α2, 116
penciclovir, 121, 203
pepsin, 70
peptomimetic, 72, 77, 78
peramivir, 181–183
peramivir structure, 232
perinatal transmission, 99
permeability, 86, 146, 151, 152, 206
Pfeiffer's bacillus, 158, 159, 161
P-glycoprotein, 74
Pgp, 74
Pharmacoenhancer, 74
pharmacokinetic booster, 74
pharmacokinetic enhancer, 80
pharmacokinetics, 61, 78, 152
Pharmasset, 145–148
phenotypic bovine diarrhea virus cellular assay, 149
phenotypic replicon assay, 139
phenotypic screen, 142
phenprocoumon–protease complex, 78
phosphate isostere, 82
phosphonate, 120–123
phosphoramidate prodrug, 124, 146, 205, 207
phosphorylation, 61, 121, 122, 145, 146, 206, 210
pibrentasvir, 140, 142, 143, 230
pibrentasvir structure, 230
Pifeltro, 68, 69
pirezenpine, 66

Pizarro, Francisco, 9–10
PK booster, 74
PK enhancer, 74
plasma-based HBV vaccine, 108
plasmid, 109, 110, 115, 165, 167, 197
pneumocystis, 38, 39
PoC, 210
polio virus, 34, 96
polycytidilic acid, 114
polyinosinic, 114
polymerase acidic protein, 184
polymerase basic proptein-1, 184
polymerase basic proptein-2, 184
polymerase chain reaction, 164, 192
potyvirus, 4
Poxviridae, 7
precipitin reaction, 101
Prezista, 79
prodrug, 65, 77, 80, 86, 118, 120–125, 144–147, 183, 185, 205, 206–210, 212, 213
proof-of-concept, 210
prophylactic agent, 196
prophylaxes, 172, 173
protease substrate, 134
protein binding, 79
protein-based vaccines, 195
ProTide, 125, 146, 147, 206
proton ion, 174
proto-oncogenes, 31
pro-viral DNA, 81
Prusof, William 57–59, 202
pyranose core structure, 180

QT prolongation, 87

radioimmunoassay, 105
raltegravir, 81, 82, 83, 223
raltegravir structure, 223
Rapivab, 181–183
RAS, 55
Rb, 55
RBD, 196

RdRp viral genes, 192
RdRp, 82, 131, 143, 184, 192, 204, 208
receptor-binding domains, 192, 196
recombinant DNA technology, 109, 110
recombinant DNA vaccine, 110
recombinant interferon, 115
recombinant vaccine, 167
Recombivax HB, 110, 111
Reed, Walter, 18
Relenza, 175–179
Rembrandt tulips, 3
remdesivir, 202–211, 233
remdesivir structure, 233
renin inhibitors, 70
replicase, 35
replicating incompetent, 201
replication-defective adenovirus, 199
replication-incompetent, 199
replicon assay, 132, 140, 146
respiratory syncytial virus, 129
retrovir, 40, 56
retroviral oncogene, 55
retrovirus, 32–36, 38, 41–46, 49, 59, 60, 66, 80, 105, 217
reverse transcriptase, 34, 35, 41, 44, 59, 60
reversible covalent inhibitors, 133–136
reversible NS3/4A inhibitors, 136–139
rhesus monkeys, 132, 208
rheumatoid arthritis, 211
ribavirin, 129, 130, 133, 138, 147, 148, 149
ribonucleic acid, 4, 33, 38, 194
ribonucleoside antiviral drugs, 203
rilpivirine, 68, 69, 90
rilpivirine structure, 221
rimantadine structure, 231
rimantadine, 174, 175
risk–benefit profile, 144

ritonavir, 72, 73, 80, 142
ritonavir structure, 222
RNA virus, 95
RNA-dependent DNA polymerase, 34, 82, 131, 143, 184, 204, 208
RNA-replication terminator, 206
rotinavir-boosting regimens, 77
Rous sarcoma virus, 26, 27, 32, 34, 36, 39, 43, 127
Rous, Peyton, 26, 30, 101, 127
RSV, 129
rule-of-five, 151, 152

salt bridge, 177
salt-retention, 215
saquinavir, 71, 72, 75–77, 79, 80, 204, 221
saquinavir structure, 221
SAR, 213
sarcoma, 26, 27, 32, 34, 36, 39, 43, 127
SARS, 189–190, 207, 208
SARS-2-CoV-2, 90, 127, 158, 184, 191
SBDD, 72, 131
scissile bond, 77
segmented RNA virus, 157
self-replicating episomal units, 109
Selzentry, 83, 88
Semper Augustus, 3
Sendai virus, 114
Ser-His-Asp catalytic triad, 138
serological test, 105
serology, 94
seropositive, 199
sero-prevalence, 200
serum hepatitis, 98
serum hepatitis antigen, 104
setrobuvir, 149
severe acute respiratory syndrome, 189–190, 207
sialic acid, 156, 175, 176, 177, 180, 182

simeprevir, 136, 137, 138, 139, 227
simeprevir structure, 227
simian immunodeficiency
    virus, 36, 45
single-stranded RNA, 191
slow bacterial pneumonia, 156
smallpox, 5, 9–13, 14, 15, 166
smallpox vaccination, 14
smallpox virus, 5, 9–13, 15
SOC, 129
social distancing, 12
sodium citrate, 102, 141, 145–
    148, 204
sofosbuvir structure, 230
solubility, 74, 76, 77, 86, 135, 136,
    151, 209
Sovaldi, 145–148, 204
Spanish influenza pandemic,
    157–160
species-specific, 113
spike protein, 70, 84, 156, 170, 188,
    191, 192, 196–200
Spiroptera carcinoma, 54
ssRNA, 191
standard of care, 129
stavudine 58, 62, 220
stavudine structure, 220
stomach cancer, 54
structurally intrinsic flexibility, 68
structure–activity relationship, 213
structure-based drug design, 72, 78,
    131, 134, 135, 177
sub-genomic amplicons, 130
substance P antagonist, 205
substrate-binding pocket, 131
substrate-like inhibitor, 133
subunit protein vaccines, 197
subunit viral vaccine, 108
super spreader, 191
supply-chain, 195
surface sialic acids, 175
surface spike glycoproteins, 193, 194

sustained viral response, 40, 67, 126
SVR, 40, 92, 126, 130
swine flu, 155, 161, 162
swine-origin H1N1, 165
Symmetrel, 173

T4 tumor cell lines, 47
Tagrisso, 41–42
Tamiflu, 179–181
Tarceva, 41
T-cell growth factor, 41, 43
T-cell inhibitor, 39
T-cell lymphoma, 35
T-cell response, 195, 198
T-cell, 28, 43, 38, 39, 41, 43, 44, 46,
    60, 61, 70, 85, 86, 89, 166, 195,
    198, 201
tegobuvir, 149
telaprevir, 133–135, 226
telaprevir structure, 226
telbivudine, 117, 118, 203, 219, 225
telbivudine structure, 219, 225
Temin, Howard, 33–35, 41
tenofovir, 62, 65, 83, 116, 117, 120–
    125, 220
tenofovir alafenamide, 83, 120,
    124, 125
tenofovir disoproxil, 65, 123, 124
tenofovir disoproxil fumarate, 65
tenofovir disoproxil structure,
    220, 225
tetanus vaccine, 166
Theiler, Max, 18, 19
thermodynamics, 71
thimerosal, 110
thioureas, 67
thumb subdomains,
    polymerases, 144
thumb-I inhibitors, 148
thymidine, 66, 219
thymidine structure, 219
thymine, 57

TIBO, 68
tipranavir, 77, 78, 222
tipranavir structure, 222
tobacco mosaic disease, 23
tobacco mosaic virus, 23, 24
toxicophore, 67
transcription, 33, 184, 194, 195
transcriptional template, 106
transfusion-associated
    hepatitis, 125
transition-state mimetics, 71, 177
transition-state structure, 177
translation, 33, 140, 194
translational machinery, 194
triamcinolone, 214
trifluorothymidine, 203, 219
trifluorothymidine structure, 219
trimeric form, 84
tripeptide, 71
triphosphate nucleoside, 209, 210
trivalent live attenuated influenza
    vaccine, 171
Trivicay, 185
trivirapine, 68
tropane bicyclic ring, 88
tropism, 46, 70, 196
trough concentrations, 74
tulip breaking virus, 4, 5
tulip mosaic virus, 4
tulipomania, 2–5
tumor suppressor gene, 55
tumor-suppressor genes, 31
typhoid bacilli, 166
typing of red blood cells, 101
tyrosine kinase inhibitor, 42
Tyzeka, 117, 203

ultraviolet-inactivated influenza
    virus, 113
UMP–CMPK, 210
undecapeptide, 134
uracil 56, 149, 209

uridine monophosphate–cytidine
    monophosphate kinase, 210

vaccines only contain the
    antigen, 167
valopicitabine, 144
van Leeuwenhoek, Antony, 1, 19
vaniprevir structure, 227
vaniprevir, 138, 227
Vaqta, 95
varicella-zoster virus, 119, 121
variegation, 2
variola major, 9
variola virus, 7
variolation, 166
vectored vaccine, 197
Veklury, 202–211
velpatasvir, 140, 141, 229
velpatasvir structure, 229
Vemlidy, 124
Victrelis, 134
Viekira Pak, 142
violent viral pneumonia, 156
Viracept, 76, 80
viral genome, 164
viral load, 85, 147, 154, 186
viral M2 ion channel envelope
    protein, 174
viral polymerase, 118
viral replication, 123, 125, 184,
    196, 204
viral replication cycle, 196
viral uncoating process, 174
viral vector vaccine, 192, 196–197
Viramune, 66
Virazole, 129
Viread, 65, 124
virion, 25, 34, 70, 84, 86, 95, 139, 156,
    167, 174, 190
viroid, 25
Viroptic, 203
virus genome integration, 81

virus interference, 112, 113
virus-related Nobel Prizes, 127
vitamin A deficiency, 54
Vitekta, 74
voxilaprevir structure, 228
voxilaprevir, 138, 228
VZV, 119

warfarin, 78
warhead, 133, 134
weakened virus, 166
Wellferon, 115
white cells, 165
WHO, 42, 105, 125, 187
wiggling and giggling, 69
World Health Organization, 42, 105, 125, 187
World Hepatitis Day, 105

Xofluza, 184–186

X-ray crystallography, 72, 131, 143, 176, 180

yeast plasmid, 110
yellow fever, 15–18
yellow fever virus, 16
yellow jack, 17

Zalcitabine, 62
zanamivir, 175–179, 231
zanamivir structure, 231
Zerit 58
zidovudine, 40, 56, 220
zidovudine structure, 220
Zika, 207
Zirgan, 121, 123
Zovirax, 120, 203
zur Hausen, Harald, 30, 31, 51, 127

α-keto amide, 133